TECHNICAL
REPORT

Outcome Measures for Effective Teamwork in Inpatient Care

Final Report

Melony E. Sorbero, Donna O. Farley,
Soeren Mattke, Susan Lovejoy

Prepared for the Agency for Healthcare Research and Quality

RAND HEALTH

This work was sponsored by the Agency for Healthcare Research and Quality (AHRQ) under contract No. 290-02-0010. The funding for this study was provided by DoD through an Interagency Agreement with AHRQ. The research was conducted in RAND Health, a division of the RAND Corporation.

Library of Congress Cataloging-in-Publication Data

Outcome measures for effective teamwork in inpatient care : final report / Melony E. Sorbero ... [et al.].
    p. cm.
    Includes bibliographical references.
    ISBN 978-0-8330-4315-3 (pbk. : alk. paper)
    1. Health care teams. 2. Outcome assessment (Medical care) 3. Medicine—Quality control. 4. Medicine—Safety measures. 5. Medical errors—Prevention. I. Sorbero, Melony E. II. Rand Corporation.
    [DNLM: 1. Outcome Assessment (Health Care) 2. Patient Care Team—organization & administration. 3. Hospitals, Military—organization & administration. 4. Inpatients. 5. Medical Errors—prevention & control. 6. Quality of Health Care. WX 162.5 O94 2008].

R729.5.H4.O98 2008
362.1068—dc22

                                                                                      2008002079

The RAND Corporation is a nonprofit research organization providing objective analysis and effective solutions that address the challenges facing the public and private sectors around the world. RAND's publications do not necessarily reflect the opinions of its research clients and sponsors.

RAND® is a registered trademark.

Published 2008 by the RAND Corporation
1776 Main Street, P.O. Box 2138, Santa Monica, CA 90407-2138
1200 South Hayes Street, Arlington, VA 22202-5050
4570 Fifth Avenue, Suite 600, Pittsburgh, PA 15213-2665
RAND URL: http://www.rand.org/
To order RAND documents or to obtain additional information, contact
Distribution Services: Telephone: (310) 451-7002;
Fax: (310) 451-6915; Email: order@rand.org

Preface

The Department of Defense (DoD) has been one of the leaders in actions to improve teamwork, which it has pursued with the goal of achieving safer care and reducing adverse events for patients served by its military hospitals. DoD and the Agency for Healthcare Research and Quality (AHRQ) have worked together to develop tools that can be used to evaluate how improved teamwork in delivering care is achieving safer outcomes for its patients. In 2004, AHRQ modified its Patient Safety Evaluation Center contract with RAND to add an analytic study to identify and test measures that have the potential to capture improvements in teamwork practices. The funding for this study was provided by DoD through an Interagency Agreement with AHRQ.

The purpose of this study was to identify a set of measures for each of three clinical areas. The measures would represent important patient safety or quality-of-care outcomes that can be expected to be affected by changes in health care teamwork effectiveness (referred to here as *teamwork-relevant outcomes*). This work addresses one step in the process of moving from teamwork training to teamwork practices that improve outcomes of care—identifying outcomes that are most likely to be affected as teamwork practices improve in an implementing organization.

This report presents the final results from the teamwork outcome measures study. Study components discussed in the report include the methods used for selecting and testing candidate measures, findings from a literature search that informed the measure-selection process, the measures identified by clinical experts as representing teamwork-sensitive outcomes, and results of the testing of a subset of these measures on the administrative data of the DoD health system.

The contents of this report will be of interest to national and state policymakers, health care organizations and clinical practitioners, and health researchers who are engaged in activities for improving the quality and safety of health care and assessing effects of those improvements on outcomes for the care process and patients.

This work was sponsored by the Agency for Healthcare Research and Quality. James B. Battles, Ph.D., served as project officer. Funding was provided by the Department of Defense. Heidi B. King, M.S., served as the project officer.

Contents

Figures

Tables

Summary

The Department of Defense (DoD) has been one of the leaders in actions to improve teamwork, which it has pursued with the goal of achieving safer care and reducing adverse events for patients served by its military hospitals. DoD and the Agency for Healthcare Research and Quality (AHRQ) have worked together to develop tools that can be used to evaluate how improved teamwork in delivering care is achieving safer outcomes for its patients. In 2004, AHRQ modified its Patient Safety Evaluation Center contract with RAND to add an analytic study to identify and test measures that have the potential to capture improvements in teamwork practices. The funding for this study was provided by DoD through an Interagency Agreement with AHRQ.

PURPOSE OF THE RESEARCH

The purpose of this project was to identify those patient safety or quality-of-care measures that could be expected to be affected by changes in teamwork effectiveness in health care (referred to here as *teamwork-relevant measures*) for three clinical areas. DoD was seeking tools to evaluate the effect on patient outcomes of improved teamwork in delivering care. This work addressed one step in the process of moving from teamwork training to teamwork practices that improve outcomes of care—the need to identify measures most likely affected as teamwork practices improve in an organization. It was an important step toward developing the capability to accurately document and assess teamwork effects.

IMPORTANCE OF TEAMWORK FOR PATIENT SAFETY

Patient safety as a problem came to widespread national attention in 2000 with the publication of the Institute of Medicine report *To Err Is Human*. Problems in communication— one of the central components of effective teamwork—have been documented as a contributing factor to both a large percentage of the sentinel events reported to the Joint Commission and to medication errors resulting in fatalities (Joint Commission, 2004a; US Pharmacopeia, 2003a,b). Clinical areas with complex and interconnected systems, such as emergency departments, surgical suites and intensive care, are at greatest risk for teamwork failures that can cause medical errors (IOM 2001). A retrospective review of malpractice claims for emergency departments suggested that appropriate teamwork might have averted more than 60 percent of the deaths and major permanent impairments that were reviewed (Risser et al., 1999).

Teamwork is the cooperative effort by individuals in a group to accomplish a common goal. Effective teamwork includes, but goes beyond, effective communication. Cannon-Bowers and colleagues (1995) identified three areas in which competencies are needed to achieve effective teamwork: teamwork knowledge, teamwork skills, and teamwork attitudes. Achieving and sustaining effective teamwork in the delivery of health care has become a patient safety priority for the U.S. health system.

CURRENT TEAMWORK IMPROVEMENT INITIATIVES

The Department of Defense (DoD) has been one of the leaders in actions to improve teamwork. It funded several medical team training initiatives across military hospitals with the goals of achieving safer care and reducing adverse events. It implemented programs modeled on

Crew Resource Management: MedTeams™, Medical Team Management (MTM), Dynamics Outcomes Management©, and, most recently, TeamSTEPPS. AHRQ made TeamSTEPPS training materials publicly available in fall 2006, through AHRQ.

A variety of health care organizations in the United States has implemented teamwork training and improvement initiatives. Presented here are some examples of key initiatives and the clinical measures they use to assess the effects of their initiatives. It is important to distinguish between the actions taken to improve teamwork (a quality-improvement process) and the effects of those improvements (which are considered in the monitoring and feedback step in the improvement process). All three organizations in the examples below are examining both changes in their teamwork processes and the effects of those changes on clinical outcomes, staff outcomes, and business outcomes:

- *VHA Inc.*, an alliance of 2,400 not-for-profit hospitals, launched a program in 20 of its hospitals in fall 2005, entitled Transformation of the Operating Room. The hospitals are implementing selected teamwork behaviors, including pre- and post-surgical briefings, and time-outs.

- *Kaiser Permanente* implemented a "Preoperative Safety Briefing" in its 30 hospitals, which it reports has reduced nursing staff turnover and wrong-site surgeries, and has increased detection of near misses (Landro, 2005).

- The *Institute for Healthcare Improvement* (IHI) is promoting teamwork through a number of its initiatives. The Improving Perinatal Care program, which is now available to hospitals throughout the United States, aims to improve care by applying evidence-based interventions and using teamwork techniques such as SBAR (Situation-Background-Assessment-Recommendation) and conflict resolution. The Transforming Care at the Bedside initiative also includes a teamwork component; the changes proposed to improve teamwork/vitality include the use of SBAR and other communication tools, enhancing collaboration and conducting multidisciplinary rounds, and enhancing teamwork among nursing staff.

It is clear from these early activities that health care organizations are identifying outcome measures that they intuitively believe are affected by teamwork. However, two issues complicate the ability to work effectively with and interpret changes in these measures. First, many initiatives assess the simultaneous effects of multiple interventions, only one of which is teamwork improvement, making it difficult to isolate the unique contribution of teamwork improvement to outcomes. Second, only limited evidence exists regarding which process and outcome measures are most directly and strongly affected by changes in teamwork effectiveness.

CONCEPTUAL FRAMEWORK

Figure S.1 provides a simple conceptual framework for the scope of work that was performed in this project. As shown in the figure, the provision of team training is expected to improve team performance, which, in turn, should result in better execution of the procedures or other activities involved in the health care being provided. These improvements in health care activities should then lead to improvements in patient outcomes—for example, fewer wrong-site surgeries or fewer cases of wrong medication dosage.

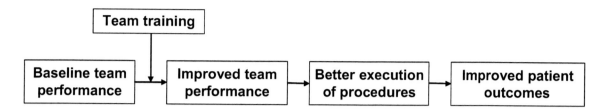

Figure S.1. Relationships Between Team Training, Teamwork Practices, and Outcomes

This project identified teamwork-relevant process and outcome measures using a multistep approach. We conducted a review of the health care literature to identify studies that assessed teamwork training and improvement activities and assessed the current state of knowledge regarding teamwork effects. Information from the literature review and other sources was used to select three clinical areas as the focus of this project. Candidate teamwork-relevant measures were identified for each of the clinical areas. These measures were rated by a clinical advisory panel with expertise in patient safety, measurement, and teamwork and the extent to which they relate to or are affected by teamwork practices. The highly rated measures were empirically tested using DoD administrative health care data. Recommendations for use of these measures were developed with the input of the clinical advisory panel, and potential areas for additional work were identified.

REVIEW OF THE LITERATURE ON TEAMWORK AND PATIENT SAFETY

The literature review focused on studies that examined teamwork effects on behaviors on the job or simulations and organizational impacts. Three categories of articles arose from our review: (1) cross-sectional studies of the effects of teamwork practices on organizational processes and patient outcomes, (2) quasi-experimental pre/post studies that examined the effects of interventions to improve teamwork and communication on care delivery processes, and (3) quasi-experimental studies that evaluated formal teamwork-training programs, which had a variety of study designs regarding pre/post measurement and use of controls. The following are our key findings for each of these three groups of studies:

- The strongest and most consistent evidence of a relationship between teamwork and patient safety comes from cross-sectional studies performed in ICUs, where adverse events occur frequently enough to detect variations. In these studies, nurses' assessments of better teamwork were related to lower risk-adjusted mortality rates and ICU length of stay for medical ICU patients, but not for surgical ICU patients.

- Studies of interventions to improve teamwork and communication generally show promising effects of improved teamwork on the quality-of-care processes. However, the interventions frequently implemented teamwork at the same time as other quality-improvement interventions, which makes it difficult to isolate the effects of the teamwork component of the intervention. Furthermore, all but one of the intervention studies used a combination of pre-/post-intervention without a control group and post-only methods to evaluate the intervention, so it was not possible to determine whether observed differences are due to the intervention of interest or to other factors that might be occurring simultaneously and affecting the outcomes examined.

- Half of the studies of teamwork training limited their evaluations to participants' attitudes toward and assessment of the training, as well as to changes in participants' understanding of the principles and skills covered in the training. Teamwork training was generally well received by participants, and some studies showed participants improved their knowledge of teamwork principles. Some studies found improvements in observer-rated behaviors and observed errors, although results are weakened by lack of controls for other factors that also might affect these outcomes. The one study that compared changes in the training group to a control group that did not receive training found significant improvements in team behaviors as a result of the teamwork training, but no improvements in clinical errors were observed.

The literature on medical teamwork is still immature, but it is developing rapidly. Virtually all of the studies found some positive relationships between teamwork or interventions and their process and/or outcome measures. From this review, we conclude that there is moderate evidence for positive relationships between teamwork implementation and patient outcomes, with the most consistent evidence being in the ICU setting. The current teamwork-training literature provides some evidence that teamwork-training programs can reduce errors in clinical practices, but they do not provide evidence on the ultimate effects on patient outcomes other than patient satisfaction.

These studies also offer insights regarding measures that could be constructed to assess effects of teamwork training on patient outcomes. A wide variety of measures was used in these studies. The most commonly used were patient satisfaction, mortality (both risk-adjusted and unadjusted), and hospital length of stay. Measures of quality of care were also used, but they varied in how quality of care was measured, ranging from patient reported to provider reported to direct observation.

IDENTIFICATION OF CLINICAL AREAS TO ADDRESS

To identify clinical areas for our study, we adapted selection criteria developed and used by the Institute of Medicine for setting action priorities for its 2003 report *Priority Areas for National Action: Transforming Health Care Quality* (Adams and Corrigan, 2003). The IOM criteria are grouped into three broad domains—impact, improvability, and inclusiveness. *Impact* encompasses the real and perceived magnitude of the problem, both to the patient and family and at the system level. *Improvability* reflects the existence of a performance gap and the possibility of narrowing it. *Inclusiveness* refers to the extent to which the clinical area encompasses a variety of types of patients and the ability to generalize findings to other clinical areas.

Using a structured approach, the RAND project team selected three clinical areas for which we would assess outcome measures. First, we created a list of candidate settings that plausibly could be included in this project from the project team's and project officer's knowledge of the existence of process or outcome measures, the literature on patient safety and quality of care, and that relevant care was delivered in at least some Military Treatment Facilities (MTFs). Next, we evaluated the candidate settings based on the decision criteria, using DoD hospital data on hospital volume and costs, the literature-review results, and the project officers' and project team's expert knowledge. These were labor and delivery, surgery, and treatment of acute myocardial infarction. The two areas of labor and delivery and surgery clearly scored highest and were included in the study. The treatment of acute myocardial infarction (AMI), from its initial presentation in the emergency room through its treatment in the hospital was included as the third clinical area because quality measures were known to exist for this area.

IDENTIFICATION, RATING, AND SELECTION OF MEASURES

A multistep process was used to select candidate teamwork-related measures for each of the three clinical areas. It included the use of a clinical advisory panel to rate the measures on their relationship to teamwork. The steps in the process are listed in Table S.1, and the actions taken in each step are described in Chapter 4.

Table S.1.
Measure-Selection Process

Step	Description
1	Developed a list of possible safety and quality process and outcome measures for labor and delivery, surgery, and treatment of acute myocardial infarction
2	Assembled a clinical advisory panel
3	Clinical advisory panel members rated each measure on its relationship to teamwork and technical soundness[a]
4	Developed a list of surviving measures, ranked by median teamwork-relatedness scores and median soundness scores from the first ratings
5	Held meeting of the clinical advisory panel to discuss the ratings
6	Clinical advisory panel members rated each surviving measure and newly suggested measures
7	Developed a final list of surviving measures, ranked by median teamwork-relatedness scores and median soundness scores from the second ratings
8	Distributed ratings and rankings to panel members for final comments and feedback

[a]Assessment of the extent to which the measure accurately captures the frequency of actual events that are occurring based on the literature or the panel's clinical judgment.

We included consideration of both patient safety and quality measures because many measures fall in a gray area that could be defined as either safety or quality, depending on the viewer's perspective. For example, complications of surgery could be the direct result of an error during the procedure or they could be due to poor-quality postsurgical follow-up. We did not want to artificially constrain the possible pool of outcomes to consider.

The clinical advisory panel performed two rounds of ratings of a comprehensive set of candidate measures identified for the three clinical areas. In each round of ratings, the panel members rated each measure according to its sensitivity to teamwork and its technical soundness. Presented in Table S.2 are summaries for each of the three clinical areas of the number of measures the clinical advisory panel rated in the first round of ratings, the number of measures added or eliminated in the rating process, and the resulting number of measures rated highly by the clinical advisory panel.

The changes to the original lists of measures varied substantially by clinical area. The labor and delivery subgroup eliminated 20 process measures and 23 outcome measures. They added five process measures and six outcome measures and suggested revisions to other measures. The surgery subgroup eliminated 29 process measures and 37 outcome measures. They also suggested the addition of 23 process measures and 22 outcome measures, which included several sets of already-established measures. The AMI subgroup eliminated five process measures and four outcome measures and added none. This result reflects the well-established set of quality and safety measures for AMI care.

Appendix C lists the measures that the advisory panel identified as having priority because of their expected sensitivity to teamwork effectiveness. This list is separated into groups of process measures and outcome measures. Within each group, we identify which measures were successfully tested with the DoD health care data; unsuccessfully tested with the DoD health care data due to lack of needed information, such as detailed pharmacy information for inpatient stays; or identified in advance as requiring information not contained in the DoD health care data. Appendix D provides descriptions of each of the measures.

Table S.2.
Number and Types of Measures Rated by Clinical Advisory Panel

	Number of Measures, by Clinical Area		
	Labor and Delivery	Surgery	AMI
Process measures			
Originally identified	22	26	32
Added in rating process	5	23	0
Eliminated in rating process	20	29	5
Resulting measures rated highly	7	20	27
Outcome measures			
Originally identified	Mother: 21; Neonate: 6	38	4
Added in rating process	Mother: 2; Neonate: 4	22	0
Eliminated in rating process	Mother: 17; Neonate: 6	37	4
Resulting measures rated highly	Mother: 6; Neonate: 4	23	0

Panelists also suggested a number of concepts for consideration at the panel meeting (Appendix E). These are clinical and operational constructs that the panelists thought likely to be related to teamwork but that have not been operationalized as measures—i.e., specifications do not exist. The panel members rated the concepts fairly highly for development and testing of new measures.

RESULTS OF EMPIRICAL TESTING OF SELECTED MEASURES

Table S.3 presents the eight highly rated outcome measures that could be constructed using the DoD health care data. None of the process measures could be constructed with the administrative data. Five labor and delivery outcome measures were successfully tested with the DoD administrative health care data. Two of the measures were for mothers and three were for neonates. Three surgery outcome measures were successfully tested with the DoD administrative health care data.

Table S.3.
Highly Rated Measures That Could Be Constructed
Using DoD Administrative Data

Labor and Delivery
Uterine Rupture
Maternal Death
Intrapartum Fetal Death
Birth Trauma, Injury to Neonate–C-Sections
Birth Trauma, Injury to Neonate–Vaginal Birth

Surgery
Failure to Rescue
Foreign Body Left in During Procedure
Mortality in Low-Mortality DRG

AMI
None

At the level of the individual hospital, occurrences of events for all the measures were too low to enable estimation of statistically significant differences in rates across years. Thus, an individual MTF could not use the measures to assess the effect of implementing an intervention to improve teamwork. At the system level, occurrences of events for only two measures were sufficiently high to test for statistically significant differences in rates over time, using data aggregated for the entire DoD health system. Specific results for each clinical area are highlighted.

For *labor and delivery* measures:

- Uterine ruptures, maternal deaths, and intrapartum fetal deaths. The frequencies of events were so low that even data pooled across all DoD hospitals were too small a sample to detect statistically significant differences over time.

- Birth trauma outcomes. Data would have to be pooled across groups of military hospitals to detect statistically significant differences over time, even for the hospitals with the largest volume of deliveries.

- Delays in obtaining complete and accurately coded records for administrative data create a risk that information on events and rates may be neither accurate nor timely.

For *surgery* measures:

- Mortality in low-mortality Diagnosis Related Groups (DRGs) and foreign body left in during surgical procedure. The frequencies of events were so low that even data pooled across all DoD hospitals were too small a sample to detect statistically significant differences over time.

- Failure to rescue. The number of patients in the denominator at individual hospitals was very small, which also results in the inability to detect significant differences, even at the system level.

- Because of the evidence that high volume is a contributor to safer surgical care, it may be appropriate to "regionalize" surgeries so that they are performed at MTFs with higher volumes.

For *care of AMI* measures:

- All of the military hospitals had small numbers of AMI discharges, so this clinical area is likely to be a lower priority for DoD to focus on for further identification and testing of teamwork-sensitive measures.

DISCUSSION AND CONCLUSIONS

The scope of work defined for this study was to identify process-of-care and outcome measures that are sensitive to teamwork for three clinical areas that the DoD, health care organizations, or other entities could use to assess the effects of teamwork-training and -improvement interventions on patient care and outcomes. The clinical areas selected were labor and delivery, surgery, and treatment for AMI. In this project, we worked with the clinical advisory panel to identify teamwork-related measures based on their clinical judgment using a modified-Delphi process, as well as published evidence. To our knowledge, this is the first study to use a systematic process to assess the teamwork relatedness of measures.

The clinical advisory panel assessed many of the candidate measures as being highly related to teamwork. Across the three clinical areas, the panel members rated 54 of the 108 candidate process measures (50 percent) and 33 of the 97 candidate outcome measures (34 percent) as highly related to teamwork.

Many of the outcome measures identified in this study and all of the process measures require information that is not available in administrative data, but that is available in medical charts.

The results of the measure-selection process and empirical testing of the highly rated measures led to a far-ranging examination of issues related to the ability to construct and effectively use teamwork-related measures. These topics fell into four areas: (1) the use of process-of-care and outcome measures, (2) data sources for construction of teamwork-related measures, (3) the application of measures of low frequency events, and (4) the need for a multidisciplinary team to identify and test teamwork-related measures. These issues are discussed briefly here and are detailed in Chapter 6.

The use of process of care and outcome measures. Process and outcome measures contribute different information for assessments of quality of care. The goals of a quality-measurement activity should drive decisions regarding which process or outcome measures should be used and how to use them.

Data sources for construction of teamwork-related measures. Various types of data could be used to assess the effects of improvements in quality or safety practices. The choice of data source is determined not only by the availability of needed data in each source but also by the available resources. Because each data source has strengths and/or weaknesses, it would be better to use measures that have been based on several data sources. The use of measures being reported for other purposes would help ensure data availability and integrity.

The application of measures of low-frequency events. The patient outcome measures tested in this study are very-low-frequency events, making them difficult to use as rates. However, a hospital could track the occurrence of events (the numerators), each of which is a serious and preventable event. When an event occurs, it would trigger an investigation of the underlying causes, and the hospital would take remedial action as part of its existing performance-improvement process.

The need for a multidisciplinary team to identify and test teamwork-related measures. To ensure the face validity[1] and technical soundness of measures identified as being teamwork-sensitive, it is important that both clinical experts and people with research and analytic skills participate in the process.

Potential Areas for Additional Work

Several specific areas for which additional work is needed were identified as this project proceeded:

- First, to explore the use of data from electronic healthcare records (EHRs), surveillance systems, and other data systems for calculating process measures for the three clinical areas in this study

- Second, to empirically test the extent to which the measures identified by the clinical advisory panel in this project are related to teamwork practices. One approach for such testing could implement teamwork-improvement interventions in a particular clinical area while measuring a range of outcome measures and the extent to which teamwork behaviors actually were being used.

- Third, to develop other measures that are considered to be sensitive to teamwork. Our advisory panel identified a number of concepts during this project that could provide a starting point for this work.

- Fourth, to identify and test relevant measures for other important clinical areas. The literature suggests that there are other clinical areas that are sensitive to teamwork—most notably the emergency department and intensive care units. A similar process could be used.

- Fifth, to assess the availability of data for measures already developed for other reporting processes. The measures reported to various entities potentially provide a rich source of additional measures for which large numbers of hospitals are collecting and reporting data.

- Sixth, to develop a comprehensive patient-safety reporting system that is designed using currently existing surveillance systems as a guide. Critical to the successful development of such systems would be identifying the range of questions the data from

---

[1] *Face validity* is the assessment that the measure appears that it will measure what it is supposed to measure.

the surveillance system would be able to address and develop standardized definitions, data-collection and data-entry methods, and audit standards and methods.

The task of identifying outcome measures that are sensitive to teamwork practices is clearly complex and involves many challenges. This study has taken an important step in embarking on that process by identifying a number of process and outcome measures that, in the judgment of our clinical advisory panel, are related to teamwork, and starting the process of testing these measures. As future work is undertaken to build on this initial study, researchers, clinicians, and policymakers can draw on not only the technical results of this study but also on the frameworks established to guide the process and to assess the numerous relevant issues that need to be addressed.

Acknowledgments

We want to thank the patient-safety, teamwork, and measurement experts who served as members of the project's clinical advisory panel. In that role, they assessed the relationships of measures to teamwork practices and shared their views on how best to use clinical process and outcome measures to assess improvements in teamwork. Their knowledgeable and thoughtful input made important contributions to this work. We also appreciate the assistance of individuals at the TRICARE Management Activity (TMA) office and their contractors at Standard Technology, Inc., who helped us obtain the needed data to ensure that our results were based on appropriate data. In addition, we thank David Tornberg, the Deputy Assistant Secretary of Defense for Clinical and Program Policies, for his support of this project.

Our work on this project also was strengthened by the support and many thoughtful comments we received throughout the course of this project from Heidi King, our TMA project officer, and Jim Battles, the AHRQ project officer of the Patient Safety Evaluation, of which this task was a part. Finally, we thank Sandra Almeida and Liz Sloss for their comments on an earlier draft of this report. Any errors of fact or interpretation are, of course, the responsibility of the authors.

Acronyms

Abbreviation	Definition
AAMC	Association of American Medical Colleges
AAP	American Association of Pediatrics
ACEI	angiotensin-converting enzyme inhibitor
ACGME	Accreditation Council for Graduate Medical Education
ACS	acute coronary syndrome
AHRQ	Agency for Healthcare Research and Quality
AIR	American Institutes for Research
AMI	acute myocardial infarction
AIO	Adverse Outcomes Index
APACHE	Acute Physiology and Chronic Health Evaluation
APSAC	anisoylated plasminogen streptokinase activator complex
ARB	Angiotensin II Receptor Blockers
CABG	coronary artery bypass graft
CAD	coronary artery disease
CMS	Centers for Medicare and Medicaid Services
CPR	cardiopulmonary resuscitation
CRM	Customer Relationship Management
CT	computerized tomography
CVA	cerebrovascular accident
CVD	cardiovascular disease
DoD	Department of Defense
DRG	Diagnosis Related Group
DVT	deep vein thrombosis
ECG	electrocardiogram
ED	emergency department
EHR	electronic healthcare record
ER	emergency room
GBS	Group B Streptococcus
HCPCS	Healthcare Common Procedure Coding System
Hg	mercury
ICD-9-CM	International Classification of Diseases, 9th Revision, *Clinical Modification Codes*
ICSI	Institute for Clinical Systems Improvement
ICU	intensive care unit
IHI	Institute for Healthcare Improvement
IOM	Institute of Medicine
IQI	Inpatient Quality Indicator
JC	the Joint Commission

LBBB	left bundle branch block
LOS	length of stay
LVEF	left ventricular ejection fraction
LVSD	left ventricular systolic dysfunction
M2	Military Health System Mart
MDC	Major Diagnostic Category
MEPRS	Medical Expense and Performance Reporting System
MTF	Military Treatment Facility
MTM	Medical Team Management
NICU	neonatal intensive care unit
NQF	National Quality Forum
NRP	Neonatal Resuscitation Program
NSQIP	National Surgical Quality Improvement Program
OR	operating room
PACE	Program for All-Inclusive Care for the Elderly
PCI	percutaneous coronary intervention
PE	pulmonary embolism
PHCT	Primary Health Care Teams
PRBC	packed red blood cells
PSI	Patient Safety Indicator
PSTT	Patient Safety Team Training
PTCA	percutaneous transluminal coronary angioplasty
RBC	red blood cells
SADR	Standard Ambulatory Data Record
SBAR	Situation-Background-Assessment-Recommendation
SCIP	Surgical Care Improvement Project
SEER	Surveillance, Epidemiology, and End Results
SIP	Surgical Infection Prevention
SIRS	Systemic Inflammatory Response Syndrome
SSI	surgical site infection
STEMI	ST-segment elevation myocardial infarction
STS	Society of Thoracic Surgeons
SYMLOG	Systemic Multiple Level Observation of Groups
UTI	urinary-tract infection
TMA	TRICARE Management Activity
VA	Veterans' Administration
VBAC	Vaginal Birth After Cesarean
VHA	Veterans' Health Administration

Chapter 1.
Introduction

The Department of Defense (DoD) has been one of the leaders in actions to improve teamwork, which it has pursued with the goal of achieving safer care and reducing adverse events for patients served by its military hospitals. DoD and the Agency for Healthcare Research and Quality (AHRQ) have worked together to develop tools that can be used to evaluate how improved teamwork in delivering care is achieving safer outcomes for its patients. In 2004, AHRQ modified its Patient Safety Evaluation Center contract with RAND to add an analytic study to identify and test measures that have the potential to capture improvements in teamwork practices.

PURPOSE OF THE RESEARCH

The purpose of this project was to identify patient safety or quality of care measures that are expected to be affected by changes in health care teamwork effectiveness (referred to here as *teamwork-relevant measures*) for three clinical areas. DoD was seeking tools to evaluate the effect on patient outcomes of improved teamwork in delivering care. This work addressed one step in the process of moving from teamwork training to teamwork practices that improves outcomes of care—the need to identify measures most likely affected as teamwork practices improve in an organization—an important step toward developing the capability to accurately document and assess teamwork effects.

IMPORTANCE OF TEAMWORK FOR PATIENT SAFETY

Patient safety as a problem came to widespread national attention with the publication in 2000 of the Institute of Medicine's report *To Err Is Human*, which estimated that 44,000 to 98,000 deaths occur annually in the United States as a result of medical errors (Kohn, Corrigan, and Donaldson, 2000). The costs of preventable adverse events, or medical errors resulting in harm, could be as high as $29 billion, with health care costs accounting for over half of the total amount (Thomas et al., 1999; Johnson et al., 1992).

Framing the Problem

Communication is one of the central components of effective teamwork. Poor communication was a contributing factor to approximately 65 percent of the sentinel events reported to the Joint Commission from 1995 through 2003, and to over 80 percent of sentinel events involving a delay in treatment (Joint Commission, 2004a). Communication problems were also the stated cause of 8 percent of medication errors reported to the MedMarx™ system in 2003 (Hicks et al., 2004); previous reports indicate that problems with communication contributed to a greater proportion of reported medication errors resulting in fatalities (US Pharmacopeia, 2003a).

Clinical areas with complex and tightly coupled systems, such as emergency departments, surgical suites, and intensive care, are at greatest risk for teamwork failures that can cause medical errors (IOM, 2001). These clinical areas are also characterized by substantial time pressure and rapidly evolving information with a high degree of ambiguity (Ostergaard, Ostergaard, and Lippert, 2004). Further complicating care, the members of teams in medicine are not static; they may even change during a procedure (Hamman, 2004). Organizations frequently identified the culture of the organization as hindering communication and teamwork. *Organizational culture* included the presence of hierarchy and intimidation, a lack of team functioning, and the established chain of communication not being followed (Joint Commission, 2004c).

A retrospective review of malpractice claims for emergency departments suggested that appropriate teamwork might have averted more than 60 percent of deaths and major permanent impairments (Risser et al., 1999). A lack of cross-monitoring[2] was identified most frequently as a primary contributor to medical errors (35 percent of cases). Other teamwork behaviors identified as primary contributors included failure to identify a protocol to use or to develop a plan (20 percent of claims), failure to prioritize tasks for the patient (22 percent of claims), and lack of assertiveness by a caregiver who raised the concern that the patient was at risk (28 percent of claims) (Risser et al., 1999). A retrospective analysis of obstetrics and gynecology–related risk-management files found that 31 percent of adverse events were associated with communication problems among team members (White et al., 2005).

What Is Teamwork?

A *team* is two or more people who interact dynamically, interdependently, and adaptively toward a common and valued goal; have specific roles or functions; and have a time-limited membership. *Teamwork* is the cooperative effort by individuals in the team to accomplish a common goal. Teamwork includes, but goes beyond, effective communication. Cannon-Bowers and colleagues (1995) identified three areas in which competencies are needed in order to achieve effective teamwork: knowledge, skills, and attitudes. *Knowledge* refers to the concepts that underlie teamwork, *skills* are techniques used to achieve effective teamwork, and *attitudes* are components of the environment and culture that make effective teamwork more likely to be achieved. Table 1.1 presents competencies in each of these areas as laid out by Baker and colleagues (2005). Achieving and sustaining effective teamwork in the delivery of health care has become a patient-safety priority for the U.S. health system. Health care organizations have invested in the development of training and implementation techniques for improving teamwork practices.

Teamwork Training

Nurses and physicians are trained separately, which results in different types of providers learning different styles of communication, and not recognizing or appreciating each other's strengths and weaknesses (Leonard. Graham, and Bonacum, 2004; IOM, 2001). Such differences may contribute to communication failures. Teamwork training is often identified as a strategy to reduce the risk of sentinel events caused by communication failures (Joint Commission, 2004c). *Teamwork training* presents information about the principles, skills and attitudes associated with effective teamwork, frequently in a multidisciplinary environment. A variety of techniques is used in teamwork training, including instructional lectures, videos, role playing, critical events training, sessions to practice skills, and simulations.

---

[2] *Cross-monitoring* is observing the actions of other team members in order to share workload and prevent medical errors (AHRQ/DoD, 2007).

With the exception of a small number of medical schools that have started to teach teamwork in their curricula, most educational programs for health care providers do not train students in teamwork (O'Connell and Pascoe, 2004). In recognition of the importance of communication and teamwork and of the lack of its inclusion in educational programs, the IOM identified approaches to develop effective teams as an area needing the attention of both the Agency for Healthcare Research and Quality (AHRQ) and private foundations (IOM, 2001). The IOM also recommended that team training for health care providers be included in patient safety programs in health care organizations (Kohn, Corrigan, and Donaldson, 2000).

National Expectations and Standards for Teamwork

Several national-level organizations have taken actions to establish expectations and standards for effective teamwork in the U.S. health care system, as follows:

- The Accreditation Council for Graduate Medical Education (ACGME) and the Association of American Medical Colleges (AAMC) recently revised the lists of competencies for physicians to include aspects of communication, coordination, and collaboration.

- The ACGME also recommended that these competencies shape resident education and be assessed periodically. Some experts suggested that teamwork competencies should be measured throughout a physician's career (Baker et al., 2005b).

- The National Quality Forum included teamwork training in its 2006 update to Safe Practices for Better Healthcare (National Quality Forum, 2006).

The Joint Commission established a goal to improve the effectiveness of communication among caregivers in its National Patient Safety Goals starting in 2004 (Joint Commission, 2004b). It also added a requirement in 2006 that health care organizations implement a standardized approach to "hand off" communications (Joint Commission, 2006).

Table 1.1
Teamwork Competencies

Competency	Definition
Knowledge competencies	
Cue/strategy associations	The linking of cues in the environment with appropriate coordination strategies.
Shared task models / situation assessment	A shared understanding of the situation and appropriate strategies for coping with task demands.
Teammate characteristics familiarity	An awareness of each teammate's task-related competencies, preferences, tendencies, strengths, and weaknesses.
Knowledge of team mission, objectives, norms, and resources	A shared understanding of a specific goal(s) or objective(s) of the team, as well as of the human and material resources required and available to achieve the objective. When change occurs, team members' knowledge must change to account for new task demands.
Task-specific responsibilities	The distribution of labor, according to team members' individual strengths and task demands.
Skill competencies	
Mutual performance monitoring	The tracking of fellow team members' efforts, to ensure that the work is being accomplished as expected and that proper procedures are followed.
Flexibility/adaptability	The ability to recognize and respond to deviations in the expected course of events, or to the needs of other team members.
Supporting/back-up behavior	The coaching and constructive criticism provided to a teammate, as a means of improving performance, when a lapse is detected or a team member is overloaded.
Team leadership	The ability to direct/coordinate team members, assess team performance, allocate tasks, motivate subordinates, plan/organize, and maintain a positive team environment.
Conflict resolution	The facility for resolving differences/disputes among teammates, without creating hostility or defensiveness.
Feedback	Observations, concerns, suggestions, and requests, communicated by team members in a clear and direct manner, without hostility or defensiveness.
Closed-loop communication/ information exchange	The initiation of a message by a sender, the receipt and acknowledgement of the message by the receiver, and the verification of the message by the initial sender.
Attitude competencies	
Team orientation (morale)	The use of coordination, evaluation, support, and task inputs from other team members to enhance individual performance and promote group unity.
Collective efficacy	The belief that the team can perform effectively as a unit, when each member is assigned specific task demands.
Shared vision	The mutually accepted and embraced attitude regarding the team's direction, goals, and mission.
Team cohesion	The collective forces that influence members to remain part of a group; an attraction to the team concept as a strategy for improved efficiency.

Mutual trust	The positive attitude that team members have for one another; the feeling, mood, or climate of the team's internal environment.
Collective orientation	The common belief that a team approach is more conducive to problem solving than an individual approach.
Importance of teamwork	The positive attitude that team members exhibit with reference to their work as a team.

Source: Baker DP, Gustafson S, Beaubien J, et al. (2005a) *Medical Teamwork and Patient Safety: The Evidence-Based Relation*. Literature Review. AHRQ Publication No. 05-0053, April 2005. Agency for Healthcare Research and Quality, Rockville MD.

CURRENT TEAMWORK-IMPROVEMENT INITIATIVES

Teamwork Activities of DoD

DoD has been one of the leaders in teamwork-improvement actions. DoD funded several medical team training initiatives across military hospitals with the goals of achieving safer care and reducing adverse events. It implemented programs modeled on Crew Resource Management: MedTeams™, Medical Team Management (MTM), Dynamic Outcomes Management©, and, most recently, TeamSTEPPS. TeamSTEPPS training materials were made available publicly in fall 2006 through AHRQ.

The American Institutes for Research (AIR) evaluated the use of these training programs, with funding by the DoD in conjunction with AHRQ. The AIR evaluation used qualitative, case-study methods to assess the quality of the programs, how well people liked the training, whether they learned what was intended, and how it affected their attitudes toward teamwork. This evaluation provided DoD with valuable information for improving the training programs.

Current Teamwork Actions by Other Health Care Organizations

A variety of health care organizations in the United States has implemented teamwork training and improvement initiatives. Presented here are some examples of key initiatives and the clinical measures they use to assess the effects of their initiatives.

We note that it is important to distinguish between actions taken to improve teamwork (a quality-improvement process) and the effects of those improvements (which are considered in the monitoring and feedback step in the improvement process). All of the organizations in these examples are examining both changes in their teamwork processes and the effects of those changes on clinical outcomes, staff outcomes, and business outcomes.

The focus of our study was on patient outcomes, which are the ultimate outcomes of interest when assessing the effects of teamwork on patient safety in the U.S. health care system. However, the other outcomes being tracked by these health care organizations also are of importance.

VHA Inc., an alliance of 2,400 not-for-profit hospitals, launched a program in 20 of its hospitals in fall 2005, entitled Transformation of the Operating Room. The hospitals are implementing selected teamwork behaviors, including pre- and post-surgical briefings, and time-outs. VHA established goals to achieve reductions in the following adverse outcomes:

- Surgical-site infections
- Adverse cardiac events
- Deep vein thrombosis

- Post-operative ventilator-associated pneumonia.

Kaiser Permanente implemented a "Preoperative Safety Briefing" in 2005 in the 30 hospitals it owns nationally. In addition to reducing nursing staff turnover, the initiation of pre-surgery briefings was credited with achieving the following clinical outcomes (Landro, 2005):

- Reduction in wrong-site surgeries
- Increase in the number of near misses detected.

The ***Institute for Healthcare Improvement*** (IHI) promotes teamwork through a number of its initiatives. Two of these are the Improving Perinatal Care program and the Transforming Care at the Bedside initiative.

The IHI Improving Perinatal Care program started in 2006 as an initiative in partnership with Ascencion Health and Premier Hospital System and involved 25 teams. The program is now available to other hospitals throughout the United States. Evidence-based interventions and teamwork techniques are implemented and chart reviews performed using IHI's Perinatal Trigger Tool to identify potentially adverse events. The program has the following four goals:

- Reduce the number of birth traumas to 3.3 adverse events per 1,000 live births
- Have maternity patients state that their preferences are known to the entire team and respected 95 percent of the time
- Improve the culture survey scores reported by perinatal units by 50 percent
- Have all medical liability claims defended because the institution's internal standards for defense were met.

IHI's Transforming Care at the Bedside initiative, which started in 2003, entered its third phase in June 2006. This initiative includes the use of communication tools, implements multidisciplinary rounds, enhances teamwork among nursing staff, and builds process-improvement competencies among front-line staff. This initiative measures the voluntary turnover rate among nurses, team development, and staff satisfaction. It also measures the following clinical measures:

- Adverse events per patient-day
- Death rates among surgical inpatients
- Death rates for non–intensive care unit, non–comfort care patients
- Number of falls per patient-day
- Occurrences of pressure ulcers per patient.

These early activities are using outcome measures they intuitively believe are affected by teamwork. However, two issues complicate the ability to interpret changes in these measures. First, many initiatives are assessing the simultaneous effects of multiple interventions, only one of which is teamwork improvement, which makes it difficult to isolate the unique contribution of teamwork improvement to improvements in the measures. Second, only limited evidence exists regarding which outcome measures are most directly and strongly affected by changes in teamwork.

This initiative contributed to filling gaps in the knowledge base by identifying clinical-process and clinical-outcome measures that are most likely to be affected as teamwork practices improve in an organization; evaluating which of these measures can be constructed using readily available administrative data; and discussing the challenges associated with the use of teamwork-relevant measures.

CONCEPTUAL FRAMEWORK

Figure 1.1 provides a simple conceptual framework for the work performed in this project. As shown in the figure, team training is expected to improve team performance, which, in turn, should result in better execution of the patient care that is provided. These improvements in health care activities should then lead to improved patient outcomes.

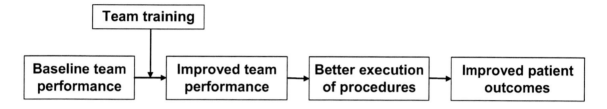

Figure 1.1. Relationships Between Team Training, Teamwork Practices, and Outcomes

This project identified teamwork-relevant process and outcome measures using a multistep approach. We conducted a review of the health care literature to identify studies that assessed teamwork training and improvement activities and assessed the current state of knowledge regarding teamwork effects (Chapter 2). Using information from the literature review and other sources, we selected three clinical areas as the focus of this project (Chapter 3). Candidate teamwork-relevant measures were identified for each of the clinical areas. These measures were rated by a clinical advisory panel with expertise in patient safety, measurement, and teamwork and the extent to which they relate to or are affected by teamwork practices (Chapter 4). The highly rated measures were empirically tested using DoD administrative health care data, and recommendations for use of these measures were developed with the input of the clinical advisory panel (Chapter 5). Chapter 6 summarizes the results of the project and discusses areas for additional work.

Chapter 2.
Review of the Literature on Teamwork and Patient Safety

Our literature review had three main purposes: to document the current state of knowledge regarding teamwork practices and their effects; to inform our selection of the three clinical areas for which measures would be identified and tested in this study; and to provide a foundation for identification of candidate teamwork-relevant outcome measures for this study. This chapter summarizes our review of the health care literature on teamwork-training and -improvement activities.

METHODS

Studies examining the relationship between teamwork, processes of care, and patient outcomes or evaluating the effects of teamwork training in medicine were identified through electronic and hand searches of the literature. Electronic searches for articles published from January 1965 through August 2006 were performed using MEDLINE, PubMed, and ABI/INFORM. Search terms included combinations of teamwork, patient safety, patient care outcomes, team training, patient care, quality of care, interdisciplinary communication, collaboration, patient care team, crisis resource management, and the names of specific teamwork-training programs, such as MedTeams.

Titles and abstracts (when available) of articles were retrieved and reviewed to determine relevance. The references of identified articles were examined for additional research not found as part of our search. In addition, recent issues of the journal *Quality and Safety in Health Care* and papers from the 2004 and 2005 AcademyHealth Annual Meetings, and the 2005 AHRQ Publication *Advances in Patient Safety: From Research to Implementation*, Volumes 1–4 were searched. For each study of interest, we abstracted key elements of the article, including author, journal, year of publication, setting, study design, aspects of teamwork examined, whether team training was included and content of curricula, process and outcome measures examined, and reported results.

We categorized articles using a framework developed by Salas and colleagues (2001) in a review of teamwork training in the airline industry. Evaluations of teamwork training and improvements may examine one or more of the following four levels of information:

1. Trainees' attitudes toward and assessment of the program

2. Trainees' understanding of the principles and skills included in the training

3. Demonstration of behaviors on the job or in simulations that are consistent with the principles and skills included in the training

4. Organizational impact, such as improved processes, reduced errors and improved outcomes.

We modified this framework to include self-reported behaviors in the workplace in the third level. Studies examining the effect of an intervention (other than teamwork training) or the relationship between teamwork and patient outcomes were included in level 4.

RESULTS

Articles describing three types of studies arose from our review:

- Cross-sectional studies of the effects of teamwork practices on organizational processes and patient outcomes, which did not include an intervention component

- Quasi-experimental pre/post studies that examined the effects of interventions to improve teamwork and communication on care delivery processes, which did not include any formal training programs or control groups

- Quasi-experimental studies that evaluated formal teamwork-training programs, which had a variety of study designs regarding pre/post measurement and use of controls.

As shown in Table 2.1, a total of 52 studies were found in the literature search. Almost half of the studies (25 studies) were teamwork-training evaluations. In identifying teamwork-training studies, we focused on those that examined information on teamwork behaviors or organizational impacts of teamwork training (levels 3 and 4 in the four levels of information).

Another 16 studies were cross-sectional studies of teamwork and related measures, and 11 studies were interventions to improve teamwork and communications. Both groups of studies examined organizational impacts (level 4 information). The studies found in each of these groups are discussed in turn below.

Table 2.1
Cross-Sectional Studies Examining Teamwork and Patient Outcomes

Type of Study	Number
Cross-sectional studies examining teamwork and outcomes	16
Interventions to improve teamwork and communications	11
Teamwork-training evaluations	25
Total number of studies	52

OUTCOME MEASURES USED BY IDENTIFIED STUDIES

Table 2.2 lists the level 3 measures (teamwork behaviors) assessed by the studies included in this review, and Table 2.3 catalogues the level 4 measures (organizational impact). The organizational-impact measures are grouped into patient-outcome, institutional-outcome, and process-of-care measures. All three types were considered in the remainder of the teamwork outcomes study as candidate measures and were rated by clinical experts.

Table 2.2
Teamwork-Behavior Measures Used in Teamwork and Teamwork-Training Studies

Teamwork-Behavior Measure	Number of Studies
Staff ratings of teamwork	11
Teamwork skills/behaviors	8
Staff ratings of communication	4
Staff ratings of collaboration	4
Task delegation/use of support personnel	2
Self-assessed performance	2
Staff ratings of organizational culture	2
Team cohesion	1
Relational coordination	1
Shared goals	1
Mutual respect	1
Patient perceptions of patient-provider interaction	1
Staff ratings of taking responsibility for patient safety	1
Self-assessed debriefing	1
Task coordination	1
Team-building skills	1
Multicultural skills	1
Role clarity	1

TYPES OF STUDIES AND FINDINGS IDENTIFIED

Collectively, the studies identified in the literature review addressed all of the framework components presented in Figure 1.1 regarding relationships between team training, teamwork practices, processes of care, and outcomes. With a few exceptions, however, the studies provided weak to moderate evidence for these relationships. We summarize here the key findings from our reviews of each of the three groups of studies identified.

Studies of Teamwork Effects on Outcomes

Sixteen cross-sectional studies were identified that examined the relationship between teamwork behaviors (level 3 measures) and patient outcomes or other measures of organizational impact (level 4 measures). Five of these studies used the ICU as the setting (Baggs et al., 1992; Baggs et al., 1999; Knaus et al., 1986; Shortell et al., 1994; Wheelan, Burchill, and Tilin, 2003). These studies used either version II or version III of the Acute Physiology and Chronic Health Evaluation (APACHE) for risk adjustment in analysis of the outcome measures (Knaus et al., 1991). Five studies focused on hospitals, but they were not limited to a specific unit; two focused on primary care teams in the outpatient setting; one focused on the operating room; one focused on patients receiving coronary artery bypass grafts; one focused on neonatal resuscitation teams; and one focused on adult day care centers participating in the Program for All-Inclusive Care for the Elderly (PACE).

The sixteen identified studies are described in Appendix A, Table A.1. The aspects of teamwork examined ranged from collaboration around the decision to transfer a patient from the ICU (Baggs et al., 1992, 1999) to multiple aspects of teamwork using composite measures

11

(Shortell et al., 1994; Wheelan, Burchill, and Tilin, 2003). All but two of the studies used staff surveys or interviews to assess the extent to which teamwork existed. Thomas and colleagues (2006) videotaped neonatal resuscitation teams and used independent observers to measure a variety of teamwork behaviors and compliance with the Neonatal Resuscitation Program (NRP) guidelines. Undre et al. (2006) also utilized an observation tool, but in the assessment of teamwork in the operating room. Nine of the studies used validated instruments or tools to assess teamwork (Boyle, 2004; Mukamel et al., 2006; Baggs et al., 1999; Meterko, Mohr, and Young, 2004; Shortell et al., 1994; Shortell et al., 2000; Thomas et al., 2006; Wheelan, Burchill, and Tilin, 2003; Campbell et al., 2001). Study results were similar, regardless of whether or not validated instruments or tools were used to assess teamwork; therefore, all studies are discussed below.

Measures studied. The five ICU studies used risk-adjusted mortality rates as an outcome measure, as did Shortell et al. (2000) in a study of outcomes for coronary artery bypass graft (CABG) surgery patients and Mukamel et al. (2006) in a study of elders participating in PACE programs. Boyle (2004) used unadjusted mortality rates. In addition to mortality rates, Baggs and colleagues (1992, 1999) also examined readmissions to the ICU. Four of the studies measured patient satisfaction (Deeter-Schmelz and Kennedy, 2003; Meterko, Mohr, and Young, 2004; Shortell et al., 2000; Campbell et al., 2001). Three studies examined length of stay (Shortell et al., 1994; Boyle, 2004; Gitell et al., 2000). Three studies examined quality of care delivered to patients, such as adherence to guidelines (Campbell et al., 2001; Undre et al., 2006; Thomas et al., 2006), whereas two examined caregiver perceptions of quality of care delivered (Deeter-Schmelz and Kennedy, 2003; Shortell et al., 1994) and two included patients' perceptions of the quality of care received (Goni, 1999; Gittel et al., 2000). Two studies assessed functional status (Shortell et al., 2000; Mukamel et al., 2006). Patient falls were assessed by two studies (Boyle, 2004; El-Jardali and Lagace, 2005).

A wide variety of other measures were examined by a single cross-sectional study. Shortell and colleagues (1994) assessed nurse turnover and ability to meet family needs. Shortell et. al. (2000) included operating time, postoperative intubation time, postoperative stroke, postoperative atrial fibrillation, and hospital cost among the measures examined. Gittell et al. (2000) measured postoperative pain. Boyle (2004) examined multiple nurse-sensitive adverse events, including urinary-tract infections, pressure ulcers, pneumonia, and failure to rescue. El-Jardali and Lagace (2005) examined nosocomial infections, tasks left undone by nurses, and medication errors. Campbell et al. (2001) examined access to care and continuity of care. Complications were assessed by Undre et al. (2006).

Table 2.3
Measures Used to Assess Effects in Teamwork and Teamwork-Training Studies

Measure	Number of Studies	Measure	Number of Studies
Patient Outcome Measures			
Patient satisfaction	8	Labor and delivery adverse outcomes	1
Risk-adjusted mortality	6	Mediastinitis	1
Unadjusted mortality	4	Patient falls	1
Risk-adjusted mortality or readmission to ICU	2	Pneumonia	1
Cardiac arrests	2	Postoperative atrial fibrillation	1
Functional health status	2	Postoperative pain and function	1
Nosocomial infections	2	Postoperative stroke	1
Adverse drug events	1	Pressure ulcers	1
Adverse events	1	Urinary incontinence	1
Complications	1	Urinary-tract infections	1
Institutional Outcome Measures			
Employee satisfaction	3	Ability to meet family member needs	1
Nurse turnover	2	Perceptions of safety climate in OR	1
Staff rating of work life quality	2	Staff morale	1
Quality and Process-of Care-Measures			
Hospital length of stay	7	Knowledge of care providers	1
Technical performance during simulations	4	Laboratory services	1
Quality of care delivered to patients	4	Medication reconciliation	1
Patient perceptions of quality of care received	4	Medication errors	1
ICU length of stay	3	Operating-room time	1
Caregiver perception of quality of care delivered	3	Postoperative intubation time	1
Hospital charges	2	Preparation of ED patients for admission	1
Hospital costs	2	Quality of health teaching	1
Access to care	1	Referrals to nurse specialists	1
Appropriate use of prophylaxis	1	Referrals to allied health workers	1
Clinical errors	1	Return to operating room	1
Compliance with guidelines	1	Service volume	1
Compliance with nutrition service recommendations	1	Surgery-related task completion	1
Comprehensiveness of care	1	Surgical count errors	1
Continuity of care	1	Tasks left undone by nurses	1
Discharge location	1	Transfers to ICU	1
Discharge plans	1	Understanding of goals for patients	1
Failure to rescue	1	Wrong-site surgeries	1

Findings on teamwork and quality and processes of care. The studies generated mixed results on the relationship between teamwork and quality of care or processes of care. Most studies did, however, find a favorable relationship between teamwork and at least one measure of quality of care or process of care examined. Two studies found that teamwork was associated with patient-perceived quality of care (Gittel et al., 2000; Goni, 1999); and two studies found associations between aspects of teamwork and provider-reported quality of care (Deeter-Schmelz and Kennedy, 2003; Shortell et al., 1994). Two studies found weak but significant relationships between aspects of teamwork and quality of care (Thomas et al. 2006; El-Jardali and Lagace, 2005). Campbell and colleagues (2001) found that team climate was associated with process quality (assessed by chart review) for diabetes, but not for asthma or angina. Three studies found that better teamwork was associated with shorter length of stay (Shortell et al., 1994; Boyle, 2004; Gitell, 2000). Undre et al. (2006) observed a positive correlation between the ratings of team communication and task completion preoperatively and postoperatively, but not intraoperatively. In general, hospital-based studies showed a more consistent relationship between teamwork and quality-of-care processes than studies done for primary care settings.

Findings on teamwork and patient outcomes. The studies generated mixed results on the relationship between teamwork and patient outcomes. Knaus et al. (1986) and Wheelan, Burchill, and Tilin (2003) found that their composite measures of teamwork were associated with lower risk-adjusted mortality for ICU patients. Baggs et al. (1992, 1999) found mixed evidence on ICU staff assessment of collaboration and risk-adjusted mortality and readmissions to the ICU. Shortell and colleagues (1994) found that caregiver interaction was negatively associated with risk-adjusted ICU length of stay and nursing turnover, and positively associated with ability to meet family member needs. It was not, however, significantly related to risk-adjusted mortality.

Studies in departments and settings other than the ICU generally found that higher ratings of teamwork were associated with statistically significantly better results for at least some of the clinical outcomes examined (Shortell et al., 2000; Gittel et al., 2000; Boyle, 2004). Two studies found a significant positive relationship between teamwork and patient satisfaction (Meterko, Mohr, and Young, 2004; Campbell et al., 2001), but two did not (Deeter-Schmelz and Kennedy, 2003; Shortell et al., 2000). Mukamel and colleagues (2006) found that team performance in a community-based program for frail-elderly individuals was significantly associated with better functional outcomes.

Assessment of Interventions to Improve Teamwork and Communication

Eleven articles described interventions to improve teamwork and communication; details on the articles are presented in Appendix A, Table A.2. Three studies evaluated the implementation of multidisciplinary daily rounds, either alone or as part of a broader restructuring of care (Young et al., 1998; Curley, McEachern, and Speroff, 1998; Uhlig et al., 2002). Leonard and colleagues (2004) reported on the introduction into the surgical care process of formalized briefings during which details of the surgery about to be performed were discussed after the patient was anesthetized. Pronovost and colleagues (2003) implemented daily goals forms as part of multidisciplinary daily rounds and as part of a multifaceted intervention to reduce catheter-related blood-stream infections. Five studies described broader initiatives to improve communication and restructure care delivery (Koerner, Cohen, and Armstrong, 1985;

Friedman and Berger, 2004; Horak et al., 2004; Mann, Marcus, and Sachs, 2006; Haig, Sutton, and Whittington, 2006).

The studies generally showed promising findings, all of the studies reporting improvements in at least some of the included measures. However, half of the studies did not report any statistical tests (Horak et al., 2004; Leonard, Graham, and Bonacum, 2004; Pronovost et al., 2003; Uhlig et al., 2002; Haig, Sutton, and Whittington, 2006; Mann, Marcus, and Sachs, 2006). Most of these studies used a combination of pre/post-intervention without a control group and post-only methods to evaluate the intervention, which makes it impossible to determine whether observed differences are due to the intervention of interest, to other activities or initiatives occurring simultaneously, or to local or national trends. One study (Koerner, Cohen, and Armstrong, 1985) included a control group, but it was a post-only evaluation; so, it was unknown whether the two groups were comparable before the intervention was implemented.

Only one study (Curley, McEachern, and Speroff, 1998) randomized patients to intervention (multidisciplinary rounds) and control groups (traditional rounds). Patients randomized to multidisciplinary rounds had lower length of stay and lower average charges than those receiving traditional rounds. There were not, however, differences in inpatient mortality or location to which patients were discharged.

Teamwork-Training Evaluations

Twenty-five articles were identified that specifically discussed teamwork-training programs. Referring to the framework in Figure 1.1, we can see that evaluations of many teamwork-training programs focused only on the first two levels—trainees' attitudes toward and assessment of the program (level 1), and trainees' understanding of the principles and skills included in the training (level 2). Only 13 of the 25 studies examined teamwork behaviors (level 3) and organizational impacts following teamwork training (level 4). None of the studies in this group examined preventable adverse patient events or other clinical outcomes, although six of them addressed measures for observed errors, technical performance, or patient satisfaction.

Findings on staff responses to teamwork training (levels 1 and 2).

Teamwork-training programs are generally well received by participants, who perceived that the training provided a unique learning experience that will help them practice more safely (Gaba et al., 1998, 2001; Flanagan, Nestel, and Joseph, 2004; Blum et al., 2004; Reznek et al., 2003; Sica et al., 1999; Holzman et al., 1995; Howard et al., 1992; O'Donnell et al., 1998; Rivers, Swain, and Nixon, 2003; O'Connell and Pascoe, 2004; Flin and Maran, 2004; Ostergaard, Ostergaard, and Lippert, 2004; Grogan et al., 2004; Stoller et al., 2004). Participants in some studies showed, and reported during follow-up interviews, improvements in their knowledge of teamwork principles, patient problem-solving skills, crisis-management skills, and communication, although the extent of improvement varied by participant characteristics, including clinical discipline and experience level (Howard et al., 1992; Dienst and Byl, 1981; Blum et al., 2004; Hope et al., 2005; Grogan et al., 2004; Awad et al., 2005).

Findings on training effects on behaviors and impacts (levels 3 and 4).

Information on the 13 studies that attempted to examine the relationships between training and subsequent behaviors and organizational impacts following teamwork training is provided in Appendix A, Table A.3. A number of studies that examined clinician behaviors in a simulated setting after training found that many teamwork inadequacies still existed. However, the lack of a pretraining assessment meant that the studies could not determine whether teamwork improvement had occurred (Gaba et al., 1998, 2001; i Gardi et al., 2001; Jacobsen et al., 2001; Howard et al., 1992). DeVita et al. (2005) found that teams performed poorly in the first simulated scenario following the didactic session, but they improved with successive simulations. Dienst and Byl (1981) reported that medical, nursing, and allied health students who were rated as providing more-comprehensive care had enhanced productivity after participating in a weekly teamwork seminar series and practicing in two-or-three person student teams.

Other studies doing pre-/post-training comparisons found improvement in observer-rated team behaviors and observed clinical errors in both simulated environments and actual practice (Sica et al., 1999; Barrett et al., 2001; Morey et al., 2002). When changes in the training group were compared to a control group that did not receive training, however, significant improvements in team behaviors were observed as a result of the teamwork training, but no improvements in clinical errors were observed (Morey et al., 2002). Clinical data suggest that teamwork training can improve the appropriate use of prophylaxis in a surgical setting (Awad et al., 2005) and reduce the number of surgical count errors (Rivers, Swain, and Nixon, 2003), although the pre/post study design does not exclude other causes for the observed improvements.

These 13 studies evaluating teamwork training have differing, sometimes conflicting, results regarding teamwork effects on behaviors or outcomes, which cannot be explained readily from the available information published in the articles. The description of teamwork and teamwork training in the studies was not sufficiently detailed to help the reader fully understand the similarities or differences in the components of teamwork examined or the aspects of teamwork included in training programs. It also has been noted that the terminology used in the teamwork-skills literature varies widely. Different names are used for the same skill, and the same name is applied to different skills, which further increases the challenge of understanding both which aspects of teamwork are most modifiable through training and which aspects are most important for improving patient outcomes (Baker et al., 2005a).

DISCUSSION

Substantial work in recent years has assessed teamwork in the delivery of medical care (Healey, Undre, and Vincent, 2004; Thomas, Sexton, and Helmreich, 2004). Aspects of teamwork identified by clinicians and experts as being important in the delivery of care include information sharing, inquiry, assertion, intention sharing, teaching, plan evaluation, workload management, vigilance and environmental awareness, leadership, communication, anticipation, coordination, redundancy, cognitive flexibility, interpersonal relations, and overall teamwork (Gaba et al., 1998; Halamek et al, 2000; Helmreich, 2000; Carthey et al., 2003; Thomas, Sexton, and Helmreich, 2004). However, as found in this review, and as also concluded by others (Thomas, Sexton, and Helmreich, 2004), there was limited evidence regarding the influence of teamwork practices on medical errors and patient outcomes.

Summary of Findings from the Literature Review

This literature review focused on studies that examined teamwork effects on behaviors on the job or in simulation (level 3) and organizational impacts (level 4). We highlight here our findings for each of the three groups of studies assessed.

- The strongest and most consistent evidence of a relationship between teamwork and patient safety comes from cross-sectional studies performed in ICUs, where the risk of mortality was substantial enough to detect variations. In these studies, higher nurses' assessments of teamwork were related to lower risk-adjusted mortality and ICU length of stay for medical ICU patients, but not for surgical ICU patients.

- Studies of interventions to improve teamwork and communication generally show promising results regarding effects of improved teamwork on quality-of-care processes. However, the interventions frequently implemented teamwork at the same time as other quality-improvement interventions, which makes it difficult to isolate the effects of the teamwork component of the intervention. Furthermore, all but one of the intervention studies used a combination of pre-/post-intervention without a control group and post-only methods to evaluate the intervention; therefore, it was not possible to determine whether observed differences are due to the intervention of interest or to other factors that might be occurring simultaneously and affecting the outcomes examined.

- Half of the studies of teamwork training limited their evaluations to participants' attitudes toward and assessment of the training, as well as to changes in participants' understanding of the principles and skills covered in the training. Teamwork training was generally well received by participants, and some studies showed that participants improved their knowledge of teamwork principles. Some studies found improvements in observer-rated behaviors and observed errors, although results are weakened by lack of controls for other factors that also might affect these outcomes. The one study that compared changes in the training group to a control group that did not receive training found significant improvements in team behaviors as a result of the teamwork training, but no improvements in clinical errors were observed.

Conclusions

The literature on medical teamwork is still immature and is developing rapidly. Virtually all of the studies found some positive relationships between teamwork or interventions and their process and/or outcome measures. The only negative result was that teamwork increased operating-room time. From our review, we conclude that there is moderate evidence for positive relationships between teamwork implementation and patient outcomes, with the most consistent evidence being in the ICU setting. The current teamwork-training literature provides some evidence that teamwork-training programs can reduce errors in clinical practices, but they do not provide evidence on ultimate effects on patient outcomes other than patient satisfaction.

These studies also offer insights regarding measures that could be constructed to assess the effects of teamwork training on patient outcomes. A wide variety of measures was used in these studies. The most commonly used were patient satisfaction, mortality (both risk-adjusted and unadjusted), and hospital length of stay. Measures of quality of care also were used, but how such quality was measured varied, ranging from patient-reported to provider-reported to direct observation.

The pathway through which teamwork-training and improvement actions ultimately affect patient outcomes can be viewed as having three basic linkage steps: (1) between teamwork training and observed improvements in teamwork behaviors, (2) between observed teamwork behaviors and self-reported teamwork, and (3) between self-reported teamwork and patient outcomes or process measures proven to be related to patient outcomes. To measure the full process of teamwork improvement from training to patient outcomes, each of these linkages can be tested empirically. It may be sufficient to demonstrate relationships at individual steps along this pathway through a variety of studies. The second linkage step is especially important to assess. Having empirical estimates of the relationship between observed behaviors and self-reported behaviors would allow researchers to use self-reported measures of behaviors in assessing outcome effects. This approach would increase the feasibility and reduce costs of measuring these relationships, rather than having to observe teamwork behaviors directly. Although there is not complete information currently, evidence is accumulating at multiple points along the pathway.

Chapter 3.
Selection of Clinical Areas

Three clinical areas for the work with measures were selected as the focus of this project. This chapter describes the process used to select the three clinical areas. It also discusses the results of the selection process.

IDENTIFICATION AND ASSESSMENT OF CLINICAL AREAS

We used a multistep approach to select the clinical areas for which we would assess process and outcome measures. First, the project team, in conjunction with the project officers, created a list of candidate clinical areas for inclusion in the study, based on the following: (1) existence of process or outcome measures, (2) patient safety or quality-of-care issues based on the team's knowledge of the literature, and (3) relevant care delivered in at least some of the medical treatment facilities (MTFs). The candidate settings were labor and delivery, cardiac ICU and medical ICU with a focus on cardiac patients, emergency department, cardiac catheters, surgery, and a subset of surgery—cardiac surgery.

We evaluated the candidate clinical areas using the selection criteria listed in Table 3.1. The criteria were adapted from those used by the Institute of Medicine for setting action priorities in its 2003 report *Priority Areas for National Action: Transforming Health Care Quality* (Adams and Corrigan, 2003). The IOM criteria were grouped into three broad domains: impact, improvability, and inclusiveness. Each clinical area was scored as high, moderate, or low by the project team on each of the criteria as shown in Table 3.2. Criteria for which no information was available were excluded from consideration.

Impact encompassed the real and perceived magnitude of the problem, both to the patient and family and at the system level. *Impact on the patient and family* consisted of the clinical burden of illness, psychological effects of potential quality problems, and patient and family dissatisfaction with quality. The clinical burden of illnesses in the clinical areas was assessed using unpublished information from the DoD project officer on hospital-service volume and costs provided through the DoD's direct-care system of MTFs for 2002. Specifically, we examined what percentage of the 100 most-frequent Diagnosis Related Groups (DRGs) across MTFs were made up by hospitalizations for the candidate clinical areas. Treatment provided through the MTFs represents a portion of the care received by military personnel, retirees, and their family members. The remainder of care is provided by the private sector through Managed Care Support Contracts. The proportion of services provided by the MTFs, which has decreased through this decade, represented 30 percent of services in 2006 (Military Health System, 2007).

The psychological effects of potential quality problems in the clinical areas were assessed according to the project team's knowledge. Ideally, we would also assess patients' and families' dissatisfaction with quality of care in the clinical area, but we did not have this information. The burden and costs of quality problems for providers and facilities were operationalized using unpublished information provided by the DoD project officer on the number and costs of malpractice cases aggregated across the MTFs in the DoD healthcare system[3] (grouped into large categories) in DoD hospitals for 1999–2004.

Improvability reflects the existence of a performance gap and the possibility of narrowing it. Three aspects of improvability were considered as part of this domain. First, the project team assessed whether there was evidence from the literature review on teamwork and teamwork training (Chapter 2) that teamwork could improve quality in the clinical area. Second, we assessed whether best practices for care were available for each of the candidate clinical areas. All of the candidate areas considered had best practices available, so this is not reported in Table 3.2 in order to simplify the table. Third, given the project team's knowledge of clinical performance measures, we assessed whether measures were available for each of the candidate clinical areas that could be used to measure variations and track improvement in quality over time.

Inclusiveness is the extent to which the clinical area encompasses a variety of types of patients and the extent to which the findings can be generalized to other clinical areas. Patient inclusiveness and condition inclusiveness were assessed in the context of the project team's knowledge of the types of conditions and patients treated in each of the candidate clinical areas. We had no information available on whether quality problems in the clinical area would disproportionately affect vulnerable populations or whether these clinical areas could be generalized to other clinical areas in MTFs.

SELECTION OF CLINICAL AREAS

Table 3.2 shows how each clinical area scored on the criteria. Labor and delivery clearly scored highest and was included in the study. Compared to the other candidate clinical areas, labor and delivery scored high on impact and moderate on improvability and inclusiveness. Deliveries were the highest-volume DRG, making up 26 percent of the 100 most common DRGs in 2002. The psychological effects of quality problems would likely be very high due to the young age of the patient involved and the emotional attachment of the parents and general expectations for a healthy infant. Labor and delivery ranked fourth in the number of acts brought in the DoD health care system. Furthermore, it ranked second in the total dollars paid in response to these acts and ranked first in dollars paid per act. The literature provided some evidence of the importance of teamwork in labor and delivery. In addition, measures existed, some of which could be constructed with administrative data. These factors led to the inclusion of labor and delivery in the study, even though the inclusiveness of the conditions treated and types of patients were moderate to low compared with the other candidate clinical areas.

Surgery was another high-scoring candidate clinical area and was included in the study. Surgeries represented approximately 20 percent of the 100 most frequent DRGs across MTFs in 2002, although many of these procedures may be concentrated at a relatively small number of MTFs. Surgery was ranked second in the number of acts brought 1999—2004 and fourth in the total dollars paid for the acts. While studies included in the literature review did not focus on surgical care, measures exist relevant to the clinical area, some of which are based on administrative data.

---

[3] The DoD refers to these cases as "acts."

Other candidate clinical areas—specifically, cardiac catheter and ICU care—were excluded because of low volume in the DoD health system. The emergency department was excluded because the magnitude of any quality issues that may exist in the DoD health system was unknown. In addition, the care delivered in emergency departments is very heterogeneous, which would make measurement challenging. The treatment of acute myocardial infarction from its initial presentation in the emergency room through its treatment in the hospital was included as the third clinical area because cardiac care was relatively common and quality measures were known to exist for this area.

Table 3.1. Criteria Used to Select Clinical Areas

Domain	Description	Operationalized
Impact Patient and family impact	Clinical burden of illness—i.e., Would care delivered in this clinical area cover common admission diagnoses?	MTF hospital volume measured as percentage of 100 most-frequent DRGs across MTFs
	Psychological effects of potential quality problems,—i.e., Would adverse events in the clinical area have a disproportionate effect on patients and their families?	Assessed from project team's knowledge.
	Patient/family dissatisfaction with quality—i.e., Are patients and their families worried about the quality of care in this clinical area?	Unable to operationalize with available data for MTFs.
System Impact	Burden of quality problems on providers and facilities—i.e., How much are MTFs and their staff concerned about quality problems in this setting and what are the costs of quality problems?	Ranking of clinical area in the number of acts brought based on care delivered in MTFs. Ranking in the total amount of money paid for acts brought and the average amount paid per act.
Improvability	Evidence that the intervention can improve quality in this clinical area—i.e., Is it plausible to assume that an intervention aimed at improving teamwork can measurably narrow the performance gap in this clinical area?	Is there evidence from the literature that improved teamwork would improve processes of care or outcomes?
	Availability of best practices of care in the clinical area—i.e., Is there sufficient agreement in the literature and the professional community about what adequate care standards encompass?	Do guidelines and best practices exist for the clinical area based on project team's knowledge? This is not reported in Table 3.2 because all candidate clinical areas had best practices available.
	Availability of measures to capture variation—i.e., Would we be able to measure deviation from accepted standards of care and to track improvement?	Do process or outcome measures exist for the clinical area, and can they be constructed with administrative data, as assessed by project team's knowledge?
Inclusiveness	Patient inclusiveness—i.e., Is care in this clinical area provided to patients with different backgrounds (race, age, sex, location, socioeconomic status)?	Project team's assessment of the inclusiveness of care provided in the clinical area.
	Equity—i.e., Would quality problems in this clinical area affect vulnerable populations disproportionately?	Unable to assess from available information.
	Condition inclusiveness,—i.e., Is care for a variety of conditions provided in this clinical area?	Project team's assessment of the inclusiveness of conditions treated in the clinical area
	Generalizability—i.e., Can lessons learned in this clinical area be applied to other clinical areas?	Unable to assess given the available information.

Table 3.2. Assessment of Potential Clinical Settings in MTFs for Development of Measures for Teamwork in Healthcare

	Labor and Delivery	Cardiac ICU & Medical ICU, focusing on cardiac patients	Emergency Department	Cardiac Catheters	OR—Cardiac Surgery	OR—Surgery
Patient and family impact						
Incidence & prevalence (MTF hospital volume)	High	Unknown	Unknown	Low	Low	High
Psychological effects of potential quality problem	Very high	High	High	High	High	High
System impact						
Burden of quality problems on providers and facilities (Are staff concerned about quality problems in setting?)	High	Unknown	Unknown	Unknown	High	High
Costs of care (malpractice)	High	Unknown	Unknown	Unknown	Medium	Medium
Improvability						
Evidence (from literature) that intervention can improve quality in this setting	Unknown	Moderate	Moderate	Unknown	Unknown	Unknown
Availability of measures to capture variation	High	High	Medium	Low	High	High
Inclusiveness						
Patient inclusiveness	Moderate	Moderate	High	Moderate	Moderate	Moderate
Condition inclusiveness	Low	Moderate	High	Low	Low	Low/Moderate

23

Chapter 4.
Measure Identification and Rating on Relationship to Teamwork

Process and outcome measures were rated by the clinical advisory panel on their relationship to teamwork and on the soundness of the measure in a two-stage process. Between the two rounds of ratings, the panel met to review the results of the first round of ratings and discuss measures with intermediate ratings. Measures could be eliminated from further consideration after each round of ratings. This chapter details the steps used to identify and rate measures for each of the three clinical areas included in the process. It also presents the results of measure ratings by the clinical advisory panel.

IDENTIFICATION, RATING, AND SELECTION OF MEASURES

A multistep process was used to select candidate teamwork-related measures for each of the three clinical areas. The process included the use of a modified-Delphi method to rate the relationship of the measures to teamwork. The method involved the clinical advisory panel independently rating the measures on their technical soundness and their relationship to teamwork, the panel meeting in person to review rating results and discuss measures with a lack of consensus, and then the panel independently re-rating the measures. The steps in the process are listed in Table 4.1, and the actions taken in each step are described below.

Table 4.1.
Measure-Selection Process

Step	Description
1	Developed a list of possible safety and quality process and outcome measures for labor and delivery, surgery, and treatment of acute myocardial infarction
2	Assembled a clinical advisory panel
3	Clinical advisory panel members rated each measure on its relationship to teamwork and technical soundness
4	Developed a list of surviving measures, ranked according to median teamwork-relatedness scores and median soundness scores from the first ratings
5	Held meeting of the clinical advisory panel to discuss the ratings
6	Clinical advisory panel members rated each surviving measure and newly suggested measures
7	Developed a final list of surviving measures, ranked according to median teamwork-relatedness scores and median soundness scores from the second ratings
8	Distributed ratings and rankings to panel members for final comments and feedback

Step 1. Developed a list of possible safety and quality process and outcome measures for labor and delivery, surgery, and treatment of acute myocardial infarction

We composed a comprehensive list of existing patient-safety, clinical process-of-care, and outcome measures developed by healthcare organizations and government agencies for each of the three clinical areas being addressed in the project. We included both patient safety and quality measures because many measures fall in a gray area that could be categorized as either safety or quality. For example, complications of surgery could be the direct result of an error during the procedure (i.e., safety) or it could be due to poor-quality post-surgical follow-up (i.e., quality). We did not want to artificially constrain the possible pool of measures to consider. In addition to established measures, we also used guidelines and literature to identify concepts[4] that could be used to develop additional measures. The actual development of measure specifications based on these concepts was not performed as part of this project. Table 4.2 lists the sources of established and potential measures. All identified measures and concepts were included on the list for assessment of their relationship to teamwork, regardless of the data sources required to construct the measure.

Table 4.2.
Primary Measure Sources

Established Measures
Joint Commission Core Measures (Joint Commission, 2007a)
Joint Commission Protocols (Joint Commission, 2007b)
Joint Commission Recommended Measures for Future Implementation (Joint Commission, 2007c)
CMS Premier Hospital Quality Indicators (CMS, 2005)
AHRQ Inpatient Quality Indicators (AHRQ, 2007a)
AHRQ Patient Safety Indicators (AHRQ, 2007b)
Society for Thoracic Surgeons/NQF Consensus Standards for Cardiac Surgery (NQF, 2007)
Veterans Health Administration Performance Measures (VA, 2007)
Institute for Clinical Systems Improvement (ICSI, 2007)
RAND
o *Quality-of-care for General Medical Conditions: A Review of the Literature and Quality Indicators* (Kerr et al., 2000a),
o *Quality-of-care for Cardiopulmonary Conditions: A Review of the Literature and Quality Indicators* (Kerr et al., 2000b),
o *Quality-of-care for Women: A Review of Selected Clinical Conditions and Quality Indicators* (McGlynn et al., 2000)
Patient Safety and Team Training in Labor and Delivery (PSTT) study measures (Mann et al., 2006)

Concepts
American Academy of Pediatrics, *Guideline for Healthy Term Newborns* (AAP, 2007)
Guidelines for Prevention of Surgical Site Infection (Mangram et al., 1999)
AHRQ Evidence Report/Technology Assessment #43, *Making Health Care Safer: A Critical Analysis of Patient Safety Practices* (Shojania et al., 2001)

---

[4] *Concepts* were potential measures that did not have technical specifications to completely define a numerator and denominator.

Step 2. Assembled a clinical advisory panel

In conjunction with DoD and AHRQ project officers, we identified a group of experts and invited them to participate on the clinical advisory panel. Experts were selected according to their experience in our three selected clinical areas (surgery, labor and delivery, or AMI), their patient-safety knowledge and experience, their familiarity and experience with teamwork concepts and training, or their experience with measure development. The 15 members of the clinical advisory panel included three registered nurses and ten physicians representing the specialties of obstetrics, surgery, anesthesiology, emergency medicine, cardiology, and pediatrics. Eight of the panel members had engaged in research or quality-improvement efforts focusing on patient safety. Five of the panel members had expertise in teamwork or team training, and three had expertise in clinical-measure development. The experts who agreed to serve on the clinical advisory panel are listed in Appendix B. The clinical advisory panel members were assigned to one of the three clinical areas based on their expertise and interests. These subgroups had primary responsibility for their clinical area for steps 3, 5, and 6, described below. Panel members were encouraged to contribute to the other clinical areas as well.

Step 3. Clinical advisory panel members rated each measure on its relationship to teamwork and technical soundness

Clinical advisory panel members were sent the measures and supporting documentation. They were asked to rate each measure on a scale of 1 to 9 on each of two dimensions: (1) the relationship of the measure to teamwork or the extent to which improvements in teamwork could improve the score on the measure and (2) the soundness of the measure in assessing the quality of care delivered. Documentation on measure validity and reliability was provided to assist panel members in this task. The ratings reflected the panel members' views of the relationship of each measure to teamwork and the measure's technical soundness, based on the evidence presented and the panel members' professional experience. Ratings of 9, 8, or 7 indicated support for the measure; ratings of 6, 5, or 4 indicated ambiguity; and ratings of 3, 2, or 1 indicated rejection. The panel members were sent measures for all three clinical areas and asked to rate measures outside their assigned clinical area if they felt comfortable doing so. They also were encouraged to suggest additional measures or concepts for the group's consideration. The results of the rating exercise, including the number of measures rated and added by the panel members, are presented later in this chapter.

Step 4. Developed a list of surviving measures, ranked by median teamwork-relatedness scores and median soundness scores from the first ratings

The measures were ranked by their median teamwork-relatedness scores and, secondarily, by their median soundness scores. Measures that were consistently rated poorly on teamwork relatedness were eliminated from further consideration. The additional measures and concepts submitted by panel members for consideration were added to the list.

Step 5. Held meeting of the clinical advisory panel to discuss the first-round ratings.

The clinical advisory panel was convened to discuss the results of the first-round ratings of the candidate measures, as well as to consider the newly proposed measures and concepts. A summary of the panelists' ratings for the three clinical areas was distributed to the members. For each measure, this summary included the median score, the average absolute difference, and the minimum and maximum scores for both teamwork-relatedness and soundness. The discussion focused on measures receiving moderate ratings on relationship with teamwork, measures for which there was disagreement among the panel members, and the newly suggested measures. The panel recommended that the measures discussed be either included for the second round of ratings or excluded from further consideration. Following the meeting, the panel members were asked to review the measures the panel had decided to exclude, to confirm their decisions.

Step 6. Clinical advisory panel members rated each surviving measure and newly suggested measures

Following the panel meeting, the clinical advisory panel performed a second rating of the measures and concepts that survived Step 5. The second round of ratings again focused on the teamwork relatedness and soundness of the measures. Several of the proposed concepts, such as "Unnecessary repetition of CT scans," were deemed by the panel members as nearly impossible to operationalize due to ambiguity about what is considered "unnecessary."

During the clinical advisory panel, members discussed two types of teamwork. The first form of teamwork took place within the care team (sharp-end). The second form referred to coordination across units in a hospital or between providers in the hospital and community (blunt-end). As part of the second round of rating, panel members were asked to indicate whether each measure reflected sharp-end or blunt-end teamwork. The following definitions were provided:

Sharp-end: Teamwork within the care team. Characterized by direct interdependence of people working closely together in the provision of care for a patient, both in terms of physical proximity and frequency of contact.

Blunt-end: Cooperation and communication across units, departments, specialty teams, or support services in the provision of care for a patient. Often characterized by lack of physical proximity in patient care and less-frequent interaction.

Step 7. Developed a final list of surviving measures based on median teamwork-relatedness scores and median soundness scores from the second ratings

The measures were ranked by their median teamwork-relatedness score from the second round of ratings and, secondarily, by their median soundness score. Measures were eliminated from further consideration if they had a median score of less than 7.0 on relationship to teamwork for AMI and surgery or a median score of less than 6.5 for labor and delivery. Measures with a median soundness score of 6 or less were removed from consideration if their median teamwork score was less than 8. Following the rating process, the concepts that were not well-defined measures were segregated for separate consideration.

Step 8. Distributed ratings and rankings to panel members for final comments and feedback

The results of the second round of ratings were distributed to the clinical advisory panel via email for comment and feedback, also via email. The summary of the panelists' ratings for the three clinical areas included for each measure the median score, the average absolute difference, and the minimum and maximum scores for both teamwork relatedness and technical soundness.

MEASURE-RATING RESULTS

Table 4.3 summarizes the number of measures the clinical advisory panel rated in the first round of ratings for each of the clinical areas. It also includes the number of additional measures and concepts that were suggested through the first round of ratings and at the panel meeting, as well as the number of measures eliminated through the rating process. The changes to the original list of measures varied substantially for the three clinical areas and are discussed below.

Table 4.3.
Number and Types of Measures Rated by Clinical Advisory Panel

	Number of Measures		
	Labor and Delivery	Surgery	AMI
Process measures			
Originally identified	22	26	32
Added in rating process	5	23	0
Eliminated in rating process	20	29	5
Resulting measures rated highly	7	20	27
Outcome measures			
Originally identified	Mother: 21; Neonate: 6	38	4
Added in rating process	Mother: 2; Neonate: 4	22	0
Eliminated in rating process	Mother: 17; Neonate: 6	37	4
Resulting measures rated highly	Mother: 6; Neonate: 4	23	0

We identified 22 process measures and 27 outcome measures for labor and delivery (21 for mothers; 6 for neonates). The labor and delivery subgroup eliminated 20 process measures and 23 outcome measures. They added five process measures and six outcome measures. They also suggested variations to existing measures and identified some measures that needed more refinement, such as the Group B Streptococcus (GBS) antibiotic timing. In addition, they discussed the possible use of a weighted outcome index that would summarize information from all outcome measures. The final list of teamwork-related measures for labor and delivery included seven process measures and ten outcome measures (six for mothers; four for neonates).

We identified 26 process measures and 38 outcome measures for surgery. The surgery subgroup eliminated 29 process measures and 37 outcome measures. They also suggested the addition of 23 process measures and 22 outcome measures, including a set of risk-adjusted outcome measures from the National Surgical Quality Improvement Program (NSQIP), the Surgical Care Improvement Project (SCIP) measures, and selected Joint Commission Surgical Infection Prevention measures (SIP 1, 3). In June 2005, CMS and the Joint Commission suspended public reporting of performance on the SIP 2 measure, Prophylactic Antibiotic Selection for Surgical Patients; therefore, this measure was removed from the list of measures to be rated. The final list of teamwork-related measures for surgery included 20 process measures and 23 outcome measures.

We identified 32 process measures and 4 outcome measures for surgery. The AMI subgroup eliminated five process measures and four outcome measures and added none. This result reflects the well-established set of quality and safety measures for AMI care. This group felt that "teamness" is reflected most strongly in processes care and that the relationship with outcomes is indirect and likely weak. Furthermore, they felt that patient characteristics and individual physician errors were more influential on outcomes. As a result, they excluded all of the candidate outcome measures. The final list of teamwork-related measures for AMI included 27 process measures and zero outcome measures.

Appendix C presents a complete list of the highly rated measures for the three clinical areas, and Appendix D describes each of the measures. Panelists also suggested a number of concepts for consideration at the panel meeting. The concepts suggested are presented in Appendix E. These are clinical and operational constructs that the panelist thought likely to be related to teamwork, but they have not been operationalized as measures (i.e., technical specifications do not exist). The panel members rated the relationship between the concepts and teamwork fairly highly and considered these were areas in which additional development and testing work could generate new measures.

Chapter 5.
Empirical Testing of Selected Measures

Measures rated by the clinical advisory panel as highly related to teamwork were empirically tested using DoD administrative health care services data. This chapter describes the methods used for this testing, presents the results of the analyses, and provides recommendations for the use of these measures.

METHODS

As a preliminary step, we assessed the data sources necessary to construct each of the highly rated measures. The data sources included administrative data, lab results, test results, nursing notes, physician notes, observation of the delivery of health care, Joint Commission data elements, and NSQIP data for surgery. The results of this assessment were reviewed and confirmed by the clinical advisory panel. We identified a subset of measures for the three clinical areas that appeared feasible to calculate using administrative data. We did not attempt to construct measures that required more than the types of data normally included in administrative data.

We tested construction of highly rated measures that appeared feasible using administrative data. To do so, we used a subset of data from the DoD military healthcare system M2 data mart[5], which captures services provided through the MTFs. We used inpatient, outpatient, and pharmacy files for 2002–2004 for care provided to patients with hospitalizations related to our three clinical areas, for all MTFs. The inpatient data used in this study included a unique patient identifier, patient age and gender, date of admission, admission status, length of stay, discharge status, up to eight procedures during the hospitalization, up to eight diagnoses, the DRG assigned to the hospital stay, MTF identifier, and, for deliveries and neonates, the method of delivery (vaginal birth or cesarean section). The outpatient data included a unique patient identifier, patient age and gender, the service date, up to four procedure codes, up to four diagnoses, the provider specialty, and an identifier for the MTF at which care was delivered. The pharmacy data included unique patient identifier, patient age, gender, issue date, days of medication supplied, number of scripts provided, national drug-code identifier, therapeutic class, and MTF identifier for prescriptions filled in the outpatient setting. Medications administered during inpatient stays were not captured in these data.

For each individual measure, the rate, we estimated number of patients eligible for the measure (denominator) and number of patients triggering the numerator of the measure, and we assessed correlations between measures for the same clinical area and for an individual measure over time. In addition, sample-size calculations were performed to determine the number of eligible patients needed to detect a 30-percent decrease from the 2002 rate at a significance level of $p=.05$ with a power of 0.80.

The clinical advisory panel was reconvened via a telephone conference to review the results of the empirical testing. During this call, the panel considered key issues related to the use

---

[5] The M2 data mart houses a variety of information including master file data on each beneficiary, as well as data on encounters for health services the beneficiaries utilized.

of the measures. The results of that discussion are incorporated in the discussion chapter of this report (Chapter 6).

RESULTS OF DATA ASSESSMENT AND EMPIRICAL TESTING

Table 5.1 summarizes the extent to which highly rated measures could be constructed using the DoD health care data. We were unable to construct any of the process measures for the three clinical areas using the DoD M2 health care data. Additional data would be required to construct these measures. The information necessary to construct some of the measures may be available from electronic data systems within individual facilities, such as detailed pharmacy data during the inpatient stay or detailed laboratory data during the inpatient stay. Other measures require information that would only be available in medical records, such as medications prescribed at discharge. We were able to construct five labor and delivery outcome measures and three surgical outcome measures.

Table 5.1.
Number and Types of Highly Rated Measures That Could Be Constructed Using DoD Administrative Data

	Labor and Delivery	Surgery	AMI
Process measures			
Successfully tested on DoD data	0	0	0
Unable to construct measures	7	20	27
Outcome measures			
Successfully tested on DoD data	Mother: 2; Neonate: 3	3	0
Unable to construct measures	Mother 4; Neonate 1	20	0

The remainder of the results are provided in three sections, one for each of the three clinical areas examined as part of the project: labor and delivery, surgery, and acute myocardial infarctions (AMIs).

Within each section are the following:

1. A list of the measures rated by the clinical advisory panel as being highly related to teamwork that were successfully tested with the DoD administrative health care data.

2. A table describing the successfully tested measures.

3. A summary of the findings from testing the measures.

4. A table with data on the rates per 1,000 for the measures successfully tested with the DoD administrative health care data.

5. A table with data on the numerators used to calculate rates for the measures successfully tested.

Additional results for Labor and Delivery and Surgery are provided in Appendixes F and G. Within the appendixes are the following:

1. A table with general information about the service activity in the clinical area for 2002–2004 for the MTFs (Tables F.1 and G.1)

2. A table with data for the denominators used to calculate the rates reported in this chapter for the measures successfully tested with the DoD administrative data (Tables F.2 and G.2).

LABOR AND DELIVERY MEASURES

Five Labor and Delivery outcome measures were successfully tested with the DoD administrative health care data. Table 5.2 describes these measures. Two of the measures were for mothers and three were for neonates:

- Maternal Measures
  o Uterine Rupture (Patient Safety Team Training [PSTT])
  o Maternal Death (PSTT)

- Neonate Measures
  o Intrapartum Fetal Death (term baby) (PSTT)
  o Birth Trauma, Injury to Neonate—C-Sections (Subset of Patient Safety Indicator [PSI] 17)
  o Birth Trauma, Injury to Neonate—Vaginal Birth (Subset of PSI 17)

It was not possible to test any of the process measures because of the lack of information in the administrative data for medications for inpatients or for times related to treatment steps.

Findings

The average rates per 1,000 for the uterine rupture, maternal death, and intrapartum fetal death measures were extremely low (Table 5.2). Reflecting these low rates, very few MTFs had patients who experienced one of these events in any given year. MTFs that had patients who experienced events had only one or two such events in a year (Table 5.4). The events occurring were not repeated in the same MTFs in the years examined.

The average rates per 1,000 for the other two measures—birth trauma and injury to neonate—were higher than those for the three measures discussed above, but the rates varied substantially across MTFs (Table 5.3). More than half of the MTFs had patients who experienced birth trauma or neonate injury (Table 5.4). MTFs that had higher rates in one year also tended to have higher rates in the other years examined (data not shown). MTFs with higher rates of birth trauma and injury for neonates delivered vaginally also tended to have higher rates for neonates delivered via cesarean section (data not shown).

To detect statistically significant changes over time (based on calculations to detect a 30-percent change from the 2002 rate), we would have had to pool data for birth trauma outcomes across MTFs. This was true even for the MTFs with the largest volume of deliveries.

We found a long time lag between when deliveries took place and when the DoD inpatient-encounter data contained complete DRG information for neonates. DRGs were used to identify neonates for this analysis. Therefore, the data for 2004 were incomplete; we were missing approximately 10,000 deliveries.

33

Table 5.2. Description of Labor and Delivery Measures Successfully Tested with DoD Administrative Health Care Data[a]

Measure	Source	Numerator	Denominator
Uterine Rupture	PSTT (Mann, Pratt, and Gluck, 2006)	Number of women who experience uterine rupture during delivery	Number of women who delivered infants
Maternal Death	PSTT (Mann, Pratt, and Gluck, 2006)	Number of women who expire as a result of giving birth	Number of women giving birth at the facility
Intrapartum Fetal Death (term baby)	PSTT (Mann, Pratt, and Gluck, 2006)	Number of intrapartum fetal deaths	Number of births at the facility
Birth Trauma, Injury to Neonate–Cesarean Section	Subset of PSI 17 (AHRQ 2007b)	Number of discharges in denominator with the *International Classification of Diseases, 9th Revision, Clinical Modification* (ICD-9-CM) code for birth trauma in any diagnosis field. **Exclude:** Infants with a subdural or cerebral hemorrhage (subgroup of birth-trauma coding) Any diagnosis code of pre-term infant (denoting birth weight of less than 2,500 grams and less than 37 weeks' gestation or 34 weeks' gestation or less). Infants with injury to skeleton and any diagnosis code of osteogenesis imperfecta.	Number of infants delivered via cesarean section
Birth Trauma, Injury to Neonate–Vaginal Delivery	Subset of PSI 17 (AHRQ, 2007b)	Number of discharges in denominator with ICD-9-CM code for birth trauma in any diagnosis field. **Exclude:** Infants with a subdural or cerebral hemorrhage (subgroup of birth trauma coding) Any diagnosis code of pre-term infant (denoting birth weight of less than 2,500 grams and less than 37 weeks' gestation or 34 weeks' gestation or less). Infants with injury to skeleton and any diagnosis code of osteogenesis imperfecta.	Number of infants delivered vaginally

[a] Detailed specifications are available from the authors.

34

Table 5.3. Outcome Rates (per 1,000) for Labor and Delivery Measures

Measure Name	2002			2003			2004		
	MTFs with Eligible Patients[a]	Rates per MTF		MTFs With Eligible Patients[a]	Rates per MTF		MTFs With Eligible Patients[a]	Rates per MTF	
		Avg.[b]	Std. Dev.[c]		Avg.[b]	Std. Dev.[c]		Avg.[b]	Std. Dev.[c]
Uterine Rupture	63	1.45	9.14	63	0.727	4.07	62	0.074	0.580
Maternal Death	64	0	0	64	0.188	1.50	62	0.010	0.080
Intrapartum Fetal Death (term baby)	70	0.016	0.136	67	0.009	0.076	65	0	0
Birth Trauma, Injury to Neonate—Cesarean Section	64	25.59	61.50	63	17.67	31.71	62	8.56	21.19
Birth Trauma, Injury to Neonate—Vaginal Delivery	64	29.14	72.72	63	17.93	39.58	63	10.91	36.74

[a] The number of MTFs with at least one patient meeting the criteria for the denominator of the measure.

[b] Among MTFs with at least one patient meeting the criteria for the denominator of the measure, the average rate per MTF for the measure.

[c] Among MTFs with at least one patient meeting the criteria for the denominator of the measure, the standard deviation for the average rate per MTF for the measure.

Table 5.4. Numerators for Labor and Delivery Outcome Measures

Measure Name	Numerator Definition	2002			2003			2004		
		MTFs with Outcomes[a]	Outcomes per MTF Avg.[b]	Std. Dev.[c]	MTFs With Outcomes[a]	Outcomes per MTF Avg.[b]	Std. Dev.[c]	MTFs with Outcomes[a]	Outcomes per MTF Avg.[b]	Std. Dev.[c]
Uterine Rupture	Number of women who experience uterine rupture during delivery	3	0.048	0.215	4	0.079	0.326	1	0.016	0.127
Maternal Death	Number of women who expire as a result of giving birth	0	0	0	1	0.016	0.125	1	0.016	0.127
Intrapartum Fetal Death (term baby)	Number of intrapartum fetal deaths	1	0.014	0.120	1	0.015	0.122	0	0	0
Birth Trauma, Injury to Neonate—Cesarean Section	Number of discharges in denominator with ICD-9-CM code for birth trauma in any diagnosis field	33	4.75	12.38	33	2.60	4.14	23	1.21	2.01
Birth Trauma, Injury to Neonate – Vaginal Delivery	Number of discharges in denominator with ICD-9-CM code for birth trauma in any diagnosis field	40	20.14	57.71	43	9.22	18.25	35	3.41	7.73

[a] Among MTFs with at least one patient meeting the denominator criteria, the average number per MTF of patients within the denominator who meet the numerator criteria for the measure.

[b] Among MTFs with at least one patient meeting the denominator criteria, the number of MTFs with at least one patient meeting the criteria for the numerator.

[c] Among MTFs with at least one patient meeting the criteria for the denominator, the standard deviation for the average number per MTF of patients within the denominator who meet the numerator criteria.

SURGERY MEASURES

Three outcome measures were successfully tested with the DoD health care data:

- Failure to Rescue (PSI 4)
- Foreign Body Left in During Procedure (PSI 5)
- Mortality in Low-Mortality DRG (PSI 2)

Descriptions of the measures are provided in Table 5.5. It was not possible to test any of the process measures because of the unavailability of information in the data for diagnostic tests or for medications for inpatients, or for times related to treatment steps.

Findings

The rates for the "failure to rescue" measure varied substantially across MTFs (Table 5.6). Approximately 40 percent of MTFs had surgical patients who triggered the failure to rescue measure (Table 5.7).The failure to rescue measure suffered, however, from small numbers of surgical patients who experienced the complications that qualified them to be included in the denominator for the measure. The MTFs with at least one patient meeting the denominator criteria had an average of 20 patients each year and a maximum of only 156 to 204 patients depending on the year (Appendix G, Table G.2). Therefore, the sample size needed to detect a 30-percent change from the average 2002 rate exceeds the total aggregate number of surgical patients eligible for the measure for all the MTFs in 2002.

The average rate per 1,000 for the measure of "foreign body left in during surgical procedure" was extremely low (Table 5.6). Reflecting these low rates, only 20–25 percent of MTFs had patients who experienced a foreign-body event in any given year. The majority of those MTFs had only one such patient, and no MTF had more than four such patients in a single year (Table 5.7). As an example, the distribution of patients experiencing these events in 2003 is tabulated as follows:

Number of patients with a foreign-body event	Number of MTFs with patients eligible for the measure (%)
0	67 (78%)
1	12 (14%)
2	4 (5%)
3	2 (2%)
4	1 (1%)

As with the failure to rescue measure, the sample size needed to detect a 30-percent change from the average 2002 rate exceeds the number of surgical patients eligible for the measure across all MTFs in 2002 combined.

The average rate per 1,000 for mortality in low-mortality DRGs also was extremely low among surgical patients (Table 5.6). Reflecting these low rates, only 5–10 percent of MTFs had patients who experienced such a death in any given year, and MTFs had at most only one or two such events (Table 5.7). The sample size needed to detect a 30-percent change from the average 2002 rate exceeds the number of surgical patients eligible for the measure across all MTFs in 2002 combined.

The rates per 1,000 for "mortality in low-mortality DRGs" and "failure to rescue" measures were negatively correlated with the number of surgery discharges at MTFs. As such, MTFs that performed fewer surgeries had higher rates for these measures (data not shown).

Table 5.5. Description of Surgery Measures Successfully Tested with DoD Administrative Health Care Data[a]

Measure	Source	Numerator	Denominator
Failure to Rescue	PSI 4 (AHRQ, 2007b)	Number of discharges in denominator with a disposition of "deceased."	Number of surgical discharges with potential complications of care listed in *failure to rescue* definition (i.e., pneumonia, DVT/PE, sepsis, acute renal failure, shock/cardiac arrest, or GI hemorrhage/acute ulcer). Exclusion criteria specific to each diagnosis. **Exclude:** Patients age 75 years and older. Neonatal patients in MDC 15. Patients transferred to an acute care facility. Patients transferred from an acute care facility. Patients admitted from a long-term care facility.
Foreign Body Left in During Procedure	PSI 5 (AHRQ, 2007b)	Number of discharges in denominator with ICD-9-CM codes for foreign body left in during procedure in any secondary diagnosis field.	Number of all surgical discharges defined by specific DRGs. **Exclude:** Patients with ICD-9-CM codes for foreign body left in during procedure in the principal diagnosis field
Mortality in Low-Mortality DRG	PSI 2 (AHRQ, 2007b)	Number of discharges in denominator with disposition of "deceased."	Number of surgical patients in DRGs with less than 0.5% mortality rate, based on NIS 1997 low-mortality DRG. If a DRG is divided into "without/with complications," both DRGs must have mortality rates below 0.5% to qualify for inclusion. **Exclude:** Patients with any code for trauma, immunocompromised state, or cancer.

[a] Detailed specifications are available from the authors.

NOTES: For these measures, the population was limited to surgical patients, although the PSI specifications are for a more general population of patients. DVT/PE = deep vein thrombosis or pulmonary embolism; GI = gastrointestinal; MDC 15 = patients with a neonatal problem.

Table 5.6. Outcome Rates (per 1,000) for Surgery Measures[a]

Measure Name	2002			2003			2004		
	MTFs with Eligible Patients[b]	Rates per MTF		MTFs with Eligible Patients[b]	Rates per MTF		MTFs with Eligible Patients[b]	Rates per MTF	
		Avg[c]	Std. Dev.[d]		Avg.[c]	Std. Dev.[d]		Avg.[c]	Std. Dev.[d]
Failure to Rescue	62	70.0	106.5	64	52.5	77.2	61	50.3	83.7
Foreign Body Left in During Procedure	88	0.295	0.780	85	0.346	1.273	82	0.453	1.635
Mortality in Low-Mortality DRG	88	0.053	0.250	85	0.199	0.764	82	0.092	0.499

[a] Note: For these measures, the population was limited to surgical patients

[b] The number of MTFs with at least one patient meeting the criteria for the denominator of the measure.

[c] Among MTFs with at least one patient meeting the criteria for the denominator of the measure, the average rate per MTF for the measure.

[d] Among MTFs with at least one patient meeting the criteria for the denominator of the measure, the standard deviation for the average rate per MTF for the measure.

Table 5.7. Numerators for Surgery Outcome Measures[a]

Measure Name	Numerator Definition	2002			2003			2004		
		MTFs with Outcomes[b]	Outcomes per MTF		MTFs with Outcomes[b]	Outcomes per MTF		MTFs with Outcomes[b]	Outcomes per MTF	
			Avg.[c]	Std. Dev.[d]		Avg.[c]	Std. Dev.[d]		Avg.[c]	Std. Dev.[d]
Failure to Rescue	Number of discharges in denominator with a disposition of "deceased."	25	2.26	4.13	24	1.88	3.71	21	2.10	4.28
Foreign Body Left in During Procedure	Number of discharges in denominator with ICD-9-CM codes for foreign body left in during procedure in any secondary diagnosis field.	18	0.318	0.720	19	0.353	0.782	18	0.305	0.661
Mortality in Low Mortality DRG	Number of discharges in denominator with a disposition of "deceased."	4	0.057	0.278	8	0.129	0.431	4	0.085	0.391

[a] For these measures, the population was limited to surgical patients.

[b] Among MTFs with at least one patient meeting the denominator criteria, the number of MTFs with at least one patient meeting the criteria for the numerator.

[c] Among MTFs with at least one patient meeting the denominator criteria, the average number per MTF of patients within the denominator that meet the numerator criteria for the measure.

[d] Among MTFs with at least one patient meeting the criteria for the denominator, the standard deviation for the average number per MTF of patients within the denominator that meet the numerator criteria.

AMI MEASURES

Four process measures were identified for possible testing using the DoD administrative data. However, none of these measures could be operationalized, for the following reasons:

- Necessary pharmacy data for inpatient stay were not available from administrative data:
  - AMI patients who received short-acting nifedipine
  - AMI patients with reduced left ventricular ejection fraction (LVEF) ($\leq$40%) or heart failure who received (contraindicated) calcium channel blocker
  - AMI patients with an LVEF$\leq$40% who received angiotension-converting enzyme (ACE) inhibitors at discharge

- Necessary laboratory data for inpatient stay were not available from administrative data:
  - AMI patients who have cholesterol assessment during hospital stay.

In addition to measurement problems, the MTFs had small numbers of AMI discharges. Only 46 to 50 MTFs had one or more AMI discharges in the 2002–2004 data with which we worked (Table 5.5). For those MTFs, the number of their annual AMI discharges ranged from 1 to 64 and their average number of AMI discharges was less than 20 (Table 5.6).

The small number of AMI discharges at any individual MTF makes it infeasible for DoD to identify and test measures sensitive to improvements in teamwork behaviors. The lack of inpatient pharmacy and laboratory data in administrative data systems might be resolved using electronic medical record data, which also might provide information for other process measures. However, given the small number of AMI discharges, this clinical area is likely to be a lower priority for the DoD than other areas with larger activity volumes.

Table 5.8.
General Information about AMI Discharges in the MTFs

	2002	2003	2004
Number of MTFs with at least one AMI discharge[a]	48	46	50
Average number of AMI discharges per MTF among MTFs with at least one AMI discharge	18.0	15.6	12.2
Standard deviation for the average number of AMI discharges per MTF	16.9	14.5	13.2
Total Number of AMI discharges	864	716	612

[a] AMI discharges were identified with an inpatient diagnosis (diagnoses 1–8) of 410 and an outpatient record for emergency treatment (Standard Ambulatory Data Record [SADR] w/ the clinic area variable = BI, BIA, BIX, BIZ) and diagnosis (diagnoses 1–4) of 410.

Chapter 6.
Discussion

The scope of work defined for this study was to identify process-of-care and outcome measures that are sensitive to teamwork for three clinical areas that the DoD, health care organizations, or other entities could use to assess the effects of teamwork-training and -improvement interventions on patient care and outcomes. The clinical areas selected were labor and delivery, surgery, and treatment for AMI. In this project, we worked with the clinical advisory panel using a modified-Delphi process, as well as published evidence, to identify teamwork-related measures based on their clinical judgment. To our knowledge, this is the first study to use a systematic process to assess the teamwork relatedness of measures. The results of the measure-selection process and the empirical testing of some of the measures led to a wide-ranging examination of issues related to the ability to construct and effectively use teamwork-related measures.

In this concluding chapter, we summarize the results of the measure rating and empirical testing. We then present a discussion that lays out the issues surrounding the use of highly rated measures that were considered by the clinical advisory panel and our project team.

SUMMARY

The clinical advisory panel rated many of the candidate measures as being highly related to teamwork. Across the three clinical areas, the panel members rated 54 of the 108 candidate process measures (50 percent) and 33 of the 97 candidate outcome measures (34 percent) as highly related to teamwork.

Many of the highly rated measures required data not typically captured in administrative health care data, including those contained in the DoD's M2 data mart. A preliminary assessment and subsequent testing of the highly rated measures determined that all the process measures and 25 of the 33 outcome measures (76 percent) required data elements not in the administrative data. The types of unavailable data frequently needed to construct the process of the measures included medications provided or results of test performed during inpatient stays, or detailed timing of care delivery. Outcome measures frequently required detailed clinical information to risk-adjust the measures.

Five outcome measures for labor and delivery, three outcome measures for surgery, and no AMI measures were successfully tested using the administrative data. Of the eight outcome measures, two (birth trauma, injury to neonate–cesarean section and birth trauma, injury to neonate–vaginal delivery) had sufficiently high rates that they could be used to monitor aggregate trends in outcomes over time for groups of MTFs. These two measures were also significantly correlated with each other, that hospitals with higher rates of birth trauma for infants born via cesarean section also had higher rates of birth trauma for infants delivered vaginally. However, these measures have some limitations, which are discussed below. The other six outcome measures had very low rates across the MTFs, and many MTFs had no events in a given year. For these measures, only very large changes in rates could be detected with statistical significance and then only if data were aggregated across all MTFs.

Two of the low-frequency surgical outcome measures, mortality in low-mortality DRGs and failure to rescue, were negatively correlated with the number of surgical discharges at the

43

MTFs; MTFs that performed fewer surgeries had higher rates for these measures than did MTFs with higher surgical volumes. This finding is consistent with other literature demonstrating a volume-outcome relationship for invasive procedures (Skinner et al., 2003; Bach et al., 2001; Taub et al., 2004). Because of the evidence of high volume as a contributor to safer surgical care, it may be appropriate to regionalize surgeries at MTFs with higher volumes.

The clinical advisory panel considered key issues related to the construction and use of teamwork-related measures during a telephone conference to review the results of the empirical testing. The topics discussed fell into five areas: (1) the use of process-of-care and outcome measures, (2) the selection of data sources for construction of teamwork-related measures, (3) the application of measures of low-frequency events, (4) the need for a multidisciplinary team to identify and test teamwork-related measures, and (5) potential areas for additional work. Each of these topics is discussed below, in turn.

USE OF PROCESS OF CARE AND OUTCOME MEASURES

Panel members were more likely to rate the candidate process measures as highly related to teamwork than the outcome measures. Process and outcome measures contribute different information for assessments of quality of care. While patients and policymakers ultimately care about the patient's outcome, many providers prefer the measurement of processes of care, which they can directly affect (Mehrotra, Bodenheimer, and Dudley, 2003). Research has demonstrated associations between many processes of care and outcomes; however, the relationships are complex. For example, patients may receive all of their clinical care appropriately and still experience a poor outcome, or they may receive poor clinical care and experience a satisfactory outcome. Furthermore, long time lags may exist between the delivery of care and resulting outcomes affected by the care.

Panel members had two main concerns about some of the candidate outcome measures: (1) that factors other than teamwork, such as unmeasured patient characteristics, were more strongly related to patient outcomes than teamwork and (2) many risk-adjustment methodologies, particularly those based on the limited clinical data available in administrative data, do not adequately account for differences between patients. The issues raised in the discussion of the relative value of process and outcome measures are relevant not only to teamwork-related measures but are frequently discussed for quality measurement more broadly. In general, the panel felt the goals of a quality-measurement activity should drive decisions regarding which process or outcome measures should be used and how to use them.

DATA SOURCES FOR CONSTRUCTION OF TEAMWORK-RELATED MEASURES

Although we focused our empirical efforts on measures that could be constructed with administrative data, various types of data could be used to assess the effects of improvements in quality or safety practices. The clinical advisory panel opined that, to effectively assess teamwork effects, it is necessary to use measures based on several types of data sources. The most common sources of data for measure construction include administrative/claims data, medical-record data, and data from surveillance systems. The choice of data source is determined not only by the availability of needed data in each source but also by the available resources. Each of these data sources has its strengths and weaknesses, which we discuss here.

Administrative /Claims Data

Administrative/claims data are frequently used to construct measures because they provide large numbers of records from multiple providers at relatively low costs. However, because administrative data were developed for purposes other than clinical monitoring—they are most frequently used for payment purposes—issues have been raised about the limited clinical information included in the data, the inability to distinguish between diagnoses present when a patient is admitted to the hospital versus those developed during the hospital stay, and the clinical accuracy of the diagnoses coded in the data.

Administrative data have the limitation of not containing the detailed clinical information needed to construct many process and outcome measures or to fully risk-adjust for differences in patient populations. Administrative data for inpatient stays use ICD-9 diagnosis codes, procedure codes, and Healthcare Common Procedure Coding System (HCPCS) codes to capture some or all of a patient's significant clinical diagnoses and the major procedures and services provided during the inpatient stay. However, the number of diagnoses and services included in the data usually are limited to a set number of codes (e.g., 10 codes), so some important codes may be omitted. Also, selection of which codes are included often is driven by what will maximize reimbursement. Moreover, whereas ICD-9 diagnosis codes capture the different conditions a patient has, they are less able to capture the variation in the severity within a disease (Iezzoni 1994), which is important for risk adjustment. The lack of ability to risk adjust contributed to the surgery and AMI subgroups of the clinical advisory panel giving lower ratings to many outcome measures based on administrative data than those outcome measures based on medical record data.

Inpatient administrative data systems typically do not capture data on common diagnostic tests and medications, or other data on the fine details of patient care, which are needed to construct measures for many inpatient processes. Furthermore, an increasing number of teamwork-sensitive process measures require data on the timing with which care is provided, which is not available in administrative data. An example is prophylactic antibiotic received within 1 hour prior to surgical incision for surgical patients.

Some outcome measures may require data not captured by inpatient administrative data systems because the measures require data on outcomes that occur outside the hospital or that involve more than one admission. Examples are 30-day mortality rates or 30-day readmission rates. A multihospital system, such as the DoD health system, could capture outcomes such as 30-day readmission rates using data for multiple hospitals in its system. However, an individual hospital will not have access to readmissions for patients who are readmitted to other hospitals.

As the above discussion indicates and as was found in this study, a set of measures constructed with administrative data is more limited than one constructed with medical records. There is evidence that using only measures that can be created with administrative data yields different quality results from a broader set of measures based on medical-record data (MacLean et al., 2006). Thus, it is possible that conclusions about the impact of teamwork training on processes of care and patient outcomes could differ depending on the types of data and measures used in evaluations.

Another substantial limitation of administrative discharge data is the current inability to distinguish between diagnoses that are present at the time of admission to the hospital and those that are developed during the hospital stay. For example, a patient may have pressure ulcers at the time of admission to the hospital or the pressure ulcers could develop during the hospital stay. Only those developed during the hospital stay should be coded as hospital-related events. Research indicates that failure to distinguish between diagnoses present on arrival and those developed in the hospital can result in the misidentification of hospitals as quality outliers (Glance et al., 2006). Capturing this additional information would require modifications to the currently used discharge forms (UB92) and would increase a hospital's overall burden. However, two states (California and New York) currently distinguish between diagnoses present at arrival and those developed during an admission, indicating that it is feasible to collect these additional data.

Coding inaccuracies can severely affect the validity of measures estimated from administrative data. For example, McCarthy et al. (2000) found that up to 30 percent of ICD-9 diagnosis codes in administrative data were not supported by information in the medical record. Frequent underreporting of complications and comorbidities in administrative data also exists (Geraci et al., 1997; Romano, Schembri, and Rainwater, 2002; Hawker et al., 1997; Faciszewski and Smith, 1995; Best et al., 2002). In addition, the average number of diagnoses codes included in hospitals' clinical data influences estimated rates for obstetric PSIs (Grobman et al., 2006) and non-obstetric complications (Romano et al., 2002; Feinglass, Koo, and Koh, 2004). In 2005, DoD National Quality Monitoring Program's Scientific Advisory Panel conducted a special study on birth trauma and injury to neonates (PSI 17). In comparing rates based on administrative data with those based on medical charts, this study found that the birth-trauma rate based on medical-record data was much lower than the rate based on administrative data. The high number of false positives with DoD administrative data indicated that the rates were higher due to coding practices rather than to true injuries to neonates. Assessments of the PSIs more generally found that the validity of coding was poor for three of the 20 provider-level PSIs and that the evidence was mixed for three indicators. The validity of coding has not been examined for six of the indicators (AHRQ, 2006).

The AHRQ Patient Safety Indicators were developed originally to provide hospitals with measurement tools for identifying and further diagnosing potential patient safety issues. AHRQ has recognized that the PSIs are being used now for other purposes, including public reporting and pay-for-performance systems, which raises more stringent performance expectations for the measures. In response to this issue, AHRQ undertook an analysis to determine which indicators could be used for these activities and under what conditions. Issues that AHRQ identified included the potential inability to adequately risk-adjust indicators for comparability across providers, potential confounding of estimates by factors not captured in the data, inconsistent use of E-codes by hospitals, and variation in coding practices and validity. Consequently, AHRQ recommends that, when the PSIs are being used for public reporting or pay-for-performance systems, hospitals be able to review their data to identify any potential coding issues and use audits or other mechanisms to ensure the accuracy and completeness of the data (Remus and Frasier, 2004).

Surveillance Systems or Data Registries

Surveillance systems or data registries represent another data source for the construction of measures. The goal of surveillance systems and data registries is to capture complete data for the universe of events of interest. Some have argued that clinical surveillance systems have the potential to produce the most precise and accurate data (Thomas and Petersen, 2003), depending on how they are designed and operated. Good surveillance systems have a number of desirable characteristics that include

- broad geographic coverage
- coverage of all patient subgroups
- regular quality-control audits
- regular training on data collection and entry
- minimal reporting burden for health care providers
- the ability to provide timely and accurate information
- processes in place to protect patient confidentiality.

One challenge of surveillance systems is that only requested data elements are collected and submitted to the system. Thus, it is critical that, early in the development of the system, developers think carefully about what questions they want to answer and how the submitted data would be used. To be effective and valid, surveillance systems must utilize standardized case definitions, data-collection methods, and data-entry methods.

Passive surveillance systems rely on the voluntary reporting of events by consumers or clinicians and, as such, tend to suffer from underreporting of the events of interest (Rosenthal and Chen, 1995). For example, adverse event reporting systems at the national, state, and local levels, are a type of passive surveillance system, and suffer from substantial underreporting of events (Thomas and Petersen, 2003).

In an active surveillance system, the entity that maintains the system works proactively with all providers and institutions responsible for reporting cases to ensure the quality and integrity of the data submitted to the system. An example of such a system is the Surveillance, Epidemiology, and End Results (SEER) program, which consists of member data registries of cancer patients (National Cancer Institute, SEER website, 2007). SEER has set a high standard for data quality for its participating data registries. It achieves and maintains that standard through a substantial infrastructure, training processes, and ongoing quality-control activities, including casefinding and recoding studies.

Medical Record Data

Medical record data are considered the gold standard for identifying adverse events (Murff et al., 2003) and constructing other quality-of-care measures. This data source, however, has its own limitations. First, abstraction of medical record data is very resource intensive and costly, requiring manual reviews to extract the relevant data for constructing the measures of interest. As a result, this data collection approach is generally considered impractical for ongoing measurement systems (Murff et al., 2003). Second, abstraction of chart data for some measures (such as some adverse events) was inconsistent, having unacceptably low inter-rater reliabilities, ranging from κ=0.40 to 0.61 (Brennan et al., 1991; Thomas, Sexton, and Helmreich, 2003).

A number of the measures based on medical records identified in this study that the clinical advisory panel rated as teamwork-related are measures that hospitals already are required to report to other organizations. For example, two of the surgery measures and seven of the AMI measures rated highly by the clinical advisory panel are reported by hospitals to The Centers for Medicare and Medicaid Services (CMS) through Medicare's Reporting Hospital Quality Data for Annual Payment Update program so that hospitals can receive their full annual payment update.

These measures are reported on the Hospital Compare website. Hospitals also report these measures to the Joint Commission.

A recent study found that hospitals report data to three to nine external organizations, the most common being CMS and the Joint Commission (Pham, Coughlan, and O'Malley, 2006), and most of the measures reported rely on medical-record data. The measures reported are generally accepted by hospitals as valid because they are evidence-based. Furthermore, some of these reporting activities include a validation process, ensuring the accuracy of the reported data. Data collected in these programs already are used for some additional purposes, such as private-sector pay-for-performance programs.

By using data already being reported by hospitals for teamwork-related measures, it should be feasible to use measures based on medical records without substantially increasing provider burden. Many of these measures are process measures that occur with adequate frequency to be able to track changes over time or make comparisons across organizations.

Many have high hopes that the use of electronic healthcare records (EHRs) will facilitate the construction of measures by making data extraction easier and faster than from medical charts. However, current EHRs frequently make use of electronic text fields rather than coded data elements and still require the manual extraction of data from the record (Persell et al., 2006). Currently, few EHRs have data fields that can be readily used to construct quality-of-care measures. In addition, new EHRs often are initially unstable, creating further delays in data extraction and management. EHRs can be developed or changed to capture specific data fields needed to populate measures, but it will be years in the future before these capabilities are likely to be widespread. Moving forward, it will be important to test the ability of EHRs to provide data for key measures. The list of process measures presented in this report offers a good starting point to test the feasibility of using EHR data for measuring levels and trends.

APPLICATION OF MEASURES OF LOW-FREQUENCY EVENTS

Numerous established quality and safety outcome measures focus on high-impact, low-frequency effects on patients, such as fetal deaths for term babies and mortality for low-mortality DRGs and other measures highly rated by the clinical advisory panel. Although low frequencies for such events are desirable, measuring and tracking such events is a challenge due to these small numbers of occurrences.

Researchers are experimenting with methods to address the limitations of low frequency events while still being able to gain valuable information from events that do occur. The appropriateness and value of each approach for measuring such outcomes would depend on the original goals for using the measures. Approaches that have been explored include

- aggregating data across many organizations
- using numerator-only measures
- using composites that combine counts for multiple measures.

Aggregation of Data Across Many Organizations

Aggregation of data across many organizations provides large enough total numbers of events for assessing trends over time, although it has the limitation of not being able to compare rates across organizations. However, some measures are too rare even for aggregation. For example, we found in this study that maternal deaths and uterine ruptures were so rare that even combining data across all DoD hospitals that perform deliveries did not generate adequate numbers of events to be able to identify statistically significant changes in rates over time. A recent study that examined trends in the rates of PSI over time across 102 hospitals in the Veterans' Administration (VA) system had to exclude two PSIs (complications of anesthesia and transfusion reaction) because of their rarity (Rosen et al., 2006).

Use of "Numerator Only" Measures to Trigger Investigation

Use of "numerator only" measures to trigger an investigation within an individual organization is an alternative approach that contributes to local quality- and safety-improvement efforts. However, it precludes being able to estimate rates because data for establishing denominators are lacking. An example of this approach is the use of *triggers*, which are key words or codes that signal the possible occurrence of errors or adverse events. Methodological work has been conducted recently to identify triggers in medical records (Rozich, Haraden, and Resar, 2003; Resar, Rozich, and Classen, 2003). Medical records are scanned to identify triggers, and, when found, a more thorough review is conducted to determine whether an adverse event did in fact occur and the severity of the event. This approach is more efficient than many customary efforts to identify adverse events.

A similar approach could be used for measures based on administrative data that have very low rates of occurrence, as was found for many of the highly rated measures tested successfully in this project. These measures could be used in a hospital's performance-improvement process by tracking occurrences of events (the numerators). Each occurrence would trigger an investigation of the underlying causes of the event, and a plan for changing practices would be developed to prevent future occurrences. Hospitals that are true learning organizations use these occurrences to make specific and organizational system changes.

Use of Composites That Combine Counts for Multiple Measures

Use of composites that combine counts for multiple measures can achieve sufficiently high rates of occurrence for the composite to be able to estimate valid rates and identify statistically significant changes in those rates over time. This approach was taken in the development of the Adverse Outcome Index (AOI) for obstetrical care, which combines ten outcome measures into a single index (Mann et al., 2006).

The development of composites involves a number of methodological issues. Here we highlight two. The first issue is how to weight the individual measures within the composite, which can affect the interpretation of results (Jacobs, Goddard, and Smith, 2005). Possible methods include weighting each measure equally (e.g., each of 10 measures receives 10 points for a total of 100 points), weighting measures by their frequency such that more-common measures drive the overall score more than less-common measures, or assigning weights based on the relative importance or severity of the individual measures. The AOI is an example of weighting by importance or severity. A second issue is that the individual measures in a composite will likely vary in the extent to which they are affected by an intervention or occurrence, such as improvements in teamwork. The inclusion of measures with a weak relationship to an intervention would dilute the overall observed relationship between the composite and the intervention. In addition, composites are still subject to the data-quality constraints of individual measures.

49

VALUE OF A MULTIDISCIPLINARY TEAM FOR IDENTIFYING AND TESTING TEAMWORK-RELATED MEASURES

The process of identifying and testing teamwork-relevant measures requires a multidisciplinary team to tackle it most effectively. To ensure the face validity and technical soundness of measures identified as being teamwork-sensitive, it is important that both clinical experts and people with research and analytic skills participate in the process. The clinical experts should represent the various professions that participate in care provision for each clinical area. We found that having physicians and nurses, both civilian and military, on our clinical panel ensured that diverse views were raised and considered in the measure-rating process. This diversity contributed to achieving face validity for the measures identified as being important for teamwork practices. The RAND team, along with the researchers serving on the panel, brought additional measurement expertise to the discussion, which enabled effective consideration of the more technical aspects of the measures and assessment of their soundness.

POTENTIAL AREAS FOR ADDITIONAL WORK

As this project proceeded, we identified several specific areas in which additional work is needed. These areas span the gamut from data development to measure specification and testing.

The measures identified and assessed in this study represent an important first step in building the capability to measure effects of improved teamwork. This project focused on identifying established measures that are most likely to be sensitive to teamwork and on assessing which of the measures could be implemented using the DoD's administrative data systems. The measures we thought could be supported by the administrative data were tested to identify issues involved in operationalizing them. Similar tests remain to be done in future work for identified measures that require other data sources, including EHRs, surveillance systems, and other administrative data systems, for calculating process measures and outcome measures for the three clinical areas in this study. For example, inpatient laboratory and medication data were not available from the DoD administrative data, but these data may be accessible through internal data systems at individual DoD hospitals.

A second area of research would address the need to empirically test the extent to which the measures identified by the clinical advisory panel in this project are related to teamwork practices. One approach for such testing could implement teamwork-improvement interventions in a particular clinical area while measuring a range of outcome measures, both before and during the intervention period. The extent to which teamwork behaviors actually were being used also would be measured, either through survey data or direct observation. Modeling techniques would be used to estimate associations between practices and process and outcome measures.

The measures used in such a study could come from several possible sources. One source is the measures identified in this study. Another approach could assess measures endorsed by NQF, which are known to be evidence-based. Alternatively, a broader set of less-established measures could be tested, such as measures from research studies that are not widely used and have not been submitted for consideration by the NQF.

A third area of future work should focus on developing other measures that are considered to be sensitive to teamwork. A number of concepts were identified by our advisory panel during this project but are not ready for use because they are not fully developed, specified, and tested.

These concepts would require additional work before they are ready to be used in an evaluative function.

Fourth, the literature suggests that other clinical areas are sensitive to teamwork—most notably, the emergency department and intensive care units. This project focused on the three clinical areas of labor and delivery, surgery, and care of patients with acute myocardial infarction, because they are core clinical areas for the DoD health system. A similar process could be used to identify and test relevant measures for other important clinical areas.

A fifth area for additional work is to assess the availability of data for measures already used for other reporting processes. As discussed above, hospitals participate in a number of reporting requirements for a variety of organizations, the most common being CMS and the Joint Commission. The measures reported to various entities provide a potentially rich source of additional measures for which large numbers of hospitals are collecting and reporting data. The clinical advisory panel rated the relationship to teamwork for some of these measures as part of this study; this relationship for measures in other clinical areas would need to be assessed.

Sixth, a comprehensive patient-safety reporting system could be developed using currently existing surveillance systems as a guide. As described above, surveillance systems are possible sources of valid data for constructing teamwork-related measures. Currently, there are no surveillance systems for monitoring and assessing a broad range of patient safety issues. Existing safety-related surveillance systems focus on hospital-acquired infections and surgical complications. Critical to the successful development of such systems would be the identification of the range of questions that the data from the surveillance system would be able to address, as well as the development of standardized definitions, data-collection, and data-entry methods, and audit standards and methods.

The task of identifying outcome measures that are sensitive to teamwork practices is clearly complex and involves numerous challenges. This study has taken an important step in that task by identifying a number of process and outcome measures that, in the judgment of our clinical advisory panel, are related to teamwork, and it has begun the process of testing those measures. As future work is undertaken to build upon this initial study, researchers, clinicians, and policymakers can draw upon not only the technical results of this study but also the frameworks established to guide the process and the assessments of numerous relevant issues that need to be addressed.

Appendix A.
Articles Included in Literature Review

Table A.1. Summary of Observational Studies of Teamwork

Author	Year	Population	Teamwork Measure	Process and Outcome Measures	Findings
Baggs et al.	1992	One ICU; 286 consecutive patients; 56 registered nurses and 31 medical residents	Collaboration in the decision to transfer patients from medical intensive care units to less-intensive care as assessed by nurses and residents on a 7-point Likert scale. Assessments were made during the shift when the patient was transferred, before the occurrence of any negative outcome.	Risk-adjusted mortality or readmission to the ICU. Risk adjustment performed using APACHE II.	Controlling for severity of illness, nurses' assessments of more collaboration were significantly associated with better patient outcomes (B=-0.22, p=.020); there was not a significant relationship with residents' assessments of collaboration (B=0.02, p=.859). Nurses' and residents' assessments of collaboration were poorly correlated (r=0.10). The predicted risk of a negative outcome was 16 percent when nurses reported no interdisciplinary collaboration in the transfer decision, but 5 percent when the decision was fully collaborative.
Baggs et al.	1999	ICU in 3 hospitals; 97 attending physicians, 63 residents, 162 staff nurses; 1432 patients	Physician-nurse collaboration in ICU with *collaboration* defined as "physicians and nurses working together, sharing responsibility for solving problems, and making decisions to formulate and carry out plans for patient care." Collaboration was assessed by nurses and physicians specifically for the decision to transfer the patient.	Risk-adjusted mortality and readmission to the ICU. Risk adjustment performed using APACHE III.	As in the preceding Baggs' work, nurses' assessments of more collaboration in the medical intensive care unit were related to better patient outcomes; physicians' assessments were not significantly related to outcomes. A one-point increase in nurses' reported collaboration on a 7-point scale was associated with a 4% decrease in the probability of a negative outcome. The predicted risk of a negative outcome was 13.9 percent when nurses reported no interdisciplinary collaboration in the transfer decision, but 3 percent when the decision was fully collaborative.

53

Table A.1—Continued

Author	Year	Population	Teamwork Measure	Process and Outcome Measures	Findings
					Nurses' and physicians' assessments of collaboration were not significantly correlated in the medical intensive care unit. Neither nurses' nor physicians' assessments of collaboration were related to patient outcomes in the surgical intensive care unit or in a community hospital ICU.
Boyle	2004	390 registered nurses on 21 medical and surgical nursing units in a 944–bed teaching hospital	Nurses completed the Nursing Work Index Revised–B instrument identifying characteristics such as practice control, nurse-physician collaboration, nurse management support, and continuity/specialization.	Nurse-sensitive adverse events, including falls, pneumonia, urinary-tract infections (UTI), pressure ulcers, cardiac arrest, mortality, failure to rescue, as well as length of stay (LOS)	Autonomy/collaboration had an inverse association with failure to rescue ($r=-0.53$) and UTIs ($r=-0.29$). Units clustered in the high group for collaboration had lower rates for pressure ulcers, falls, pneumonia, and death and shorter LOS than the group of units with low scores for collaboration.
Campbell et al.	2001	A stratified random sample of 60 general practices in six areas of England	A team-climate inventory was completed by all staff employed by 48 of the practices. The survey assesses how people work together, how frequently they interact, whether they have identified aims and objectives, and how much practical support and assistance are given.	Data from medical records were extracted to identify aspects of care identified as necessary for three chronic conditions: asthma in adults, angina, and Type 2 diabetes. Data were also collected from the appropriate health authority on rates of use for preventive care for each practice. A survey of a random sample of patients within each practice was used to assess access to care, continuity of care,	Team climate, as reported by staff, was associated with quality of care for diabetes, access to care, continuity of care and overall satisfaction. Team climate was not significantly associated with quality of care for asthma or angina.

Table A.1—Continued

Author	Year	Population	Teamwork Measure	Process and Outcome Measures	Findings
				Interpersonal aspects of care, and patient's overall satisfaction.	
Deeter-Schmelz, Kennedy	2003	68 patient-care teams, including nurses and multiskilled practitioners in a large regional medical center; 1,598 patients	Team members completed a 5-item questionnaire measuring team cohesion	Quality-of-care as measured by 5-item questionnaire completed by team members; patient satisfaction measured by 9-item instrument.	A strong relationship between team cohesion and quality of patient care was demonstrated, but no support was found for a direct relationship between cohesion and patient satisfaction.
El-Jardali, Lagace	2005	7,108 nurses from 80 acute care hospitals in Ontario, Alberta, and British Columbia were surveyed	Survey data were extracted for the following variables: staffing, support services, teamwork, pleasant work environment, tasks left undone, and adverse events. Nurses were asked to indicate on a 4-point scale whether they have a lot of teamwork with physicians and whether collaboration exists between nurses and physicians.	The survey included questions about the frequency of occurrence of adverse events for patients under their care, including medication errors, nosocomial infections, and patient falls with injuries. An additional variable is tasks left undone at the end of a shift.	Teamwork was shown to have a weak but significant relationship to tasks left undone, which is predictive of adverse events. A direct relationship between teamwork and adverse events was not examined.
Gitell et al.	2000	9 hospitals; 333 providers, including physicians, nurses, physical therapists, social workers and case managers; 878 patients	Providers completed a cross-sectional questionnaire assessing strength of communication and relationship ties between providers.	Quality-of-care as measured by a 15-item patient questionnaire; length of stay, postoperative pain, and function assessed through patient questionnaire.	Quality-of-care was significantly improved by relational coordination ($p<.001$). Postoperative pain was significantly reduced by relational coordination ($p=.041$), and postoperative functioning was significantly improved by frequency of of shared goals ($p=.035$), and the degree

Table A.1—Continued

Author	Year	Population	Teamwork Measure	Process and Outcome Measures	Findings
					of mutual respect ($p=.001$) among caregivers. Length of stay was significantly shortened (53.77%, $p=.001$) by relational coordination.
Goni	1999	31 Primary Health Care Teams (PHCTs) made up of 600 medical and nonmedical professionals offering health services in Navarre, Spain	Team members completed a questionnaire (42.6% response rate) regarding goals, empowerment, relationship and communication, flexibility, and recognition and appreciation.	Team performance was measured for 3 different stakeholders: (1)Administration (efficiency, economy, efficacy) (2) Users (index of quality based on reliability, responsiveness, courtesy, empathy, and knowledge of the patient as assessed by a random sample of patients) (3) Workers (satisfaction).	The author found a positive relationship between team design variables and some individual variables and a positive relationship between the team design and some of the variables that assess performance. For example, team performance provided higher levels of satisfaction, quality, and efficacy. No significant relationship was found between team performance and economy or efficiency.
Knaus et al.	1986	19 ICUs in 13 hospitals; 5,030 patients	Interaction and coordination between medical and nursing staff in the ICU as reported in a questionnaire completed by either the unit's medical or nursing director.	ICU risk-adjusted mortality rate. Risk adjustment performed using APACHE II.	Higher levels of interaction and coordination between medical and nursing teams were related to lower risk-adjusted mortality rates, controlling for other clinical characteristics. This was a greater factor than the unit's administrative structure, the extent to which specialized treatments were used, or the hospital's teaching status. Effect size and level of significance were not included in the article.

Table A.1—Continued

Author	Year	Population	Teamwork Measure	Process and Outcome Measures	Findings
Meterko, Mohr, and Young	2004	125 VA Hospitals	8,454 hospital employees (52% response rate) completed the Zammuto and Krakower culture questionnaire indicating whether the organization's culture type is teamwork, entrepreneurial, bureaucratic, or rational.	Patient-satisfaction data from a VHA national database	Teamwork culture was positively associated with inpatient satisfaction and had the strongest relationship among the culture types.
Mukamel et al.	2006	26 Sites for the Program for All-Inclusive Care for the Elderly (PACE), including 3,401 PACE enrollees	1,209 team members (65% response rate) completed a 7-question PACE team-performance survey.	Deterioration in functional status, as measured by increasing limitation of activities of daily living (ADLs), urinary incontinence, and mortality	Multivariate regression models with risk adjustment showed that better team performance was associated with greater improvement in the ability to perform ADLS at 3 months ($p<.05$) and at 12 months ($p<.01$); less deterioration in urinary incontinence at 12 months ($p<.05$); and no difference in mortality within one year of enrollment.
Shortell et al.	1994	42 ICUs, 17,440 patients	Caregiver interaction, which was a composite measure of assessed culture, leadership, communication, coordination, and problem solving/conflict management was assessed by surveying ICU staff.	Risk-adjusted mortality rates, risk-adjusted length of stay (using APACHE III), nurse turnover. Also evaluated technical quality of care and ability to meet family-member needs by surveying nurses, physicians, residents, and ward clerks and secretaries.	Caregiver interaction was negatively associated with risk-adjusted ICU length of stay ($B=-0.34, p\leq.05$) and nursing turnover ($B=-0.36, p\leq.05$), and positively associated with evaluated technical quality of care ($B=0.81, p\leq.01$) and higher-evaluated ability to meet family-member needs ($B=0.74, p\leq.01$), but was not significantly related to risk-adjusted mortality.

Table A.1—Continued

Author	Year	Population	Teamwork Measure	Process and Outcome Measures	Findings
Shortell et al.	2000	3,045 CABG patients; 16 hospitals; 54 clinicians including surgeons, anesthesiologists, nurses, technicians	Clinicians completed a 20-item questionnaire assessing the organizational culture. A group culture emphasizes affiliation, teamwork, coordination, and participation.	Risk-adjusted outcome (including death, return to OR, postoperative stroke, mediastinitis, or post-op atrial fibrillation); functional status (SF36), clinical efficiency (length of stay>10 days, OR time, postoperative intubation time); cost; patient satisfaction	A supportive group culture was associated with shorter postoperative intubation times (p=.01) and higher patient physical (p=.005) and mental (p=.01) functional health status scores 6 months after CABG but longer operating room times (p=.004).
Thomas et al.	2006	132 neonatal resuscitation teams caring for infants born by cesarean section at a 500-bed urban teaching hospital, with an 80-bed NICU and approximately 3,400 deliveries performed.	Neonatal resuscitation teams were videotaped, and two independent observers measured a variety of teamwork behaviors used to develop three components: communication (information sharing and inquiry), management (workload management and vigilance), and leadership (assertion, intentions shared, evaluation of plans, leadership).	Two nurses not involved in assessing teamwork rated noncompliance with item in the Neonatal Resuscitation Program (NRP) guidelines (not performed or performed incorrectly) and assessed overall quality of the resuscitation.	Interrater agreement was either fair or good for all individual teamwork behaviors except workload management, vigilance, and leadership, for which it was slight. Communication was correlated with nurses' ratings of overall resuscitation quality (Spearman p=-0.236, p=.007), total noncompliance with all NRP steps administered (Spearman p=-0.214, p=.014), as well as noncompliance during NRP preparation and initial steps (Spearman p=-0.230, p=.008). Management was significantly correlated with total noncompliance with all NRP steps administered (Spearman p=-0.201, p=.021), and noncompliance during NRP preparation and initial steps (Spearman p=-0.252, p=.003), but not nurses' ratings of overall resuscitation quality. Leadership was

Table A.1—Continued

Author	Year	Population	Teamwork Measure	Process and Outcome Measures	Findings
					significantly correlated with nurses' ratings of overall resuscitation quality (Spearman $p=-0.288, p<.001$), but not noncompliance with NRP steps.
Undre et al.	2006	50 operations in general surgery in a single operating theater	Teamwork behaviors including communication, cooperation, coordination, shared leadership, and monitoring were observed and rated by a post-doc psychologist.	Surgical-task completion, as observed by a surgical observer, and complications. Tasks were grouped into three categories (patient, equipment, and communication) and three stages (pre-op, post-op, and op).	Ratings of team communication were positively correlated with overall surgical-task completion in both the pre-op and post-op phases ($r=0.478$, $p<.001$ and $r=.345, p=.007$, respectively). Team coordination ratings were positively correlated with equipment-task completions during the post-op phase ($r=.321. p=.01$). There were no correlations between team-performance ratings and task completion during the operative phase, and there were no significant relationships between task completion and the occurrence of complications.
Wheelan, Burchill, and Tilin	2003	17 ICUs in 9 hospitals; 394 staff, including registered nurses (75%), physicians, unit Clerks, and unlicensed assistive personnel	ICU staff completed the Group Development Questionnaire, which assesses group dependency, conflict, trust and structure, and work and productivity.	Risk-adjusted mortality. Risk adjustment performed using APACHE III	Higher group-development ratings were correlated with lower standardized mortality ratios ($r=-0.662, p=.004$).

Table A.2. Summary of Studies of Interventions to Improve Teamwork and Communication

Author	Year	Population	Intervention	Outcome Measures	Findings
Curley, McEachern, and Speroff	1998	1,102 patients on inpatient medical service at MetroHealth Medical Center, a 742-bed, acute care county hospital. Patients were excluded if they were transferred to another service or if less than 50% of their stay was on the medical floor.	Patients were randomized to medical services with traditional rounds (n=535) or medical services with multidisciplinary rounds (n=567). Traditional rounds were conducted with physicians only, with medical orders written throughout the day. Weekly social service rounds were held with social work, nutrition, and interns. Multidisciplinary rounds included physicians, nurse, pharmacist, nutritionist, and social worker, with orders written during the rounds; no weekly social service rounds were needed.	Length of stay, total hospital charges, hospital deaths, discharges home, discharges to skilled/interim care facility, provider satisfaction, ancillary service efficiency, and patient perception of quality of care	Bivariate analyses showed patients randomized to multidisciplinary rounds had lower length of stay than those receiving traditional rounds (5.46 versus 6.06, p=.0006) and lower average total charges ($6,681 versus $8,090, p=.002), with no significant differences in the percentage of patients dying in the hospital, discharged home, or discharged to skilled/interim care facilities. The differences in length of stay and total charges remained significant in multivariate analyses. Provider-satisfaction surveys indicated that the multidisciplinary-rounds group had a greater understanding of patient care, more effective communication and more teamwork than the traditional-rounds group (p<.006 for each). Multidisciplinary rounds resulted in more-appropriate use of orders for administration of aerosols, approaching statistical significance (91.7% of orders deemed appropriate versus 73.6% of orders in traditional rounds; p=.075). Nutrition services' recommendations were more likely to be implemented with multidisciplinary rounds than traditional rounds (100% versus 80%; p=.018).
Friedman and Berger	2004	General surgery in a tertiary care hospital	Restructuring consisted of defining team-member roles; removing redundancies in roles; communicating better; initiating daily meetings between senior residents, case managers and charge nurses; and	Risk-adjusted length of stay and patient satisfaction. Risk adjustment was performed using DRG-based mean case weights, patient age, gender, and source of admission.	The mean length of stay decreased by 1.18 days for patients on the private service and 0.89 days for patients on the ward service (p<.001 for both wards when using a log scale to reduce the effect of outliers). In addition, following the restructuring, over 80% of general surgery patients rated the quality of care during their hospital stay as very good or good. Satisfaction data were not collected prior to the restructuring

Table A.2—Continued

Author	Year	Population	Intervention	Outcome Measures	Findings
			conducting monthly meetings with the entire patient care team.		of general surgery teams.
Haig, Sutton, and Whittington.	2006	All staff in a comprehensive-care medical center	A multifaceted effort to increase the use of Situation, Background, Assessment, Recommendation (SBAR) included training, pocket cards for clinicians, "cheat sheets" at telephones, inclusion of SBAR components in reporting documents, etc.	Medication reconciliation, adverse patient events, and adverse drug events	Medication reconciliation at patient admission increased from a mean of 72% to a mean of 88%, and discharge reconciliation improved from a mean of 53 percent to a mean of 89 percent. The rate of adverse patient events, as measured by a monthly random sample of 20 patient charts, was reduced from a baseline of 89.9 per 1,000 patient-days in October 2004 to 39.96 per 1,000 patient-days overall in the following year. Adverse drug events decreased from a baseline of 29.97 per 1,000 patient days to 17.64 per 1,000 patient-days. No statistical tests were reported.
Horak et al.	2004	A medical unit at George Washington University Hospital with an average census of 39 patients.	Two team-building meetings attended by nurses and physicians; the creation of ground rules for collaboration; reinstituting monthly meeting between nurses and physician, with a focus on building consensus on issues and solutions; semi-weekly discharge-planning meetings attended by a nurse, social worker, dietitian, physical therapist, and the entire medical team; the posting of pictures of all members of nursing	Staff ratings of the environment at the time of the survey (post intervention) and "before now," including patient care; nurse-physician collaboration; problem solving; unit procedures and nurse and physician morale; pharmacy, laboratory, radiology, dietary, transport, supplies, computers, and space.	On a 5-point scale, both nurses and physicians reported improved patient care; nurse-physician collaboration; problem solving; unit procedures and nurse and physician morale; pharmacy, laboratory, supplies, and computers. No statistical tests were reported.

Table A.2—Continued

Author	Year	Population	Intervention	Outcome Measures	Findings
			and house staff assigned to the unit during the month; the use of a dry-erase board as a communications center; a comprehensive orientation to the unit for medical students, interns, and residents; labels on charts indicating physician and nurses responsible for patient's care.		
Koerner, Cohen, and Armstrong	1985	Two 27-bed medical units (1 intervention, 1 control); 180 patients from the intervention ward; 100 patients from the control ward completed survey. Medical records for 118 intervention and 116 control patients reviewed for outcomes.	A demonstration project of a collaborative practice. Established a nursing practice through collaboration with physicians.	Patient perceptions of patient-provider interaction, quality of care received, knowledge of care providers, quality of health teaching, and quality of environment. From chart review: average length of stay, average number of laboratory tests ordered, number of cardiac arrests, number of deaths, number of transfers to ICU, number of referrals to nurse specialists, referrals to	There were no differences in the patient outcomes. Patients in the collaborative-practice unit reported better provider-patient interaction (e.g., identify doctor by appearance, how completely nurses/doctors answered questions, patient involvement by nurses in their care plan), knowledge of providers (e.g., patient-rated knowledge and skill of primary doctor, patient-rated knowledge and skill of primary nurse), quality of care (e.g., care received planned with you in mind), nurses do much to help you feel better), health teaching (e.g., know why medication given, receive explanation of treatments and tests, given enough information to care for self), and some environment measures (e.g., needs for privacy met, treated with respect on unit, treated with respect in hospital) than did control patients. No multivariate analyses were performed, and no corrections were made to significance levels for multiple testing.

Table A.2—Continued

Author	Year	Population	Intervention	Outcome Measures	Findings
				allied health workers, number of discharge plans	
Leonard, Graham, and Bonacum and Defontes and Surbida	2004	Orange County Kaiser surgical program	Formalized briefings were introduced into the surgical-care process after the patient was anesthetized. Briefing-tool development was a collaborative effort of surgeons, anesthetists, operating room nurses, technicians, managers.	Wrong-site surgeries; nursing turnover; employee satisfaction; perceptions of safety climate in the OR; ratings of teamwork, communication, taking responsibility for patient safety	Wrong-site surgeries did not occur after briefings were implemented. Nursing turnover decreased by 16%, employee satisfaction increased by 19%, and perceptions of a safety climate increased from good to outstanding. Improvements were also reported in ratings of teamwork climate, communication, feeling responsible for patient safety, and reports of medical errors being handled appropriately. Specific data and statistical tests not reported.
Mann, Marcus, and Sachs	2006	Beth Israel Deaconess Medical Center Labor and Delivery unit	Deaconess implemented a teamwork initiative as part of a larger effort to increase patient safety.	Adverse outcomes for high-risk premature births and term deliveries were measured using an Adverse Outcome Index (AOI). The AOI is the percentage of patients with one or more of 10 identified adverse events. Additionally, malpractice claims and suits were measured over a three-year period.	The AOI score for high-risk premature infants improved 47%, term deliveries improved 14%, and there was a 16% overall improvement. Malpractice claims and suits decreased by more than 50% over the three-year period. No statistical tests were reported.
Pronovost et al.	2003	Johns Hopkins Hospital 16-bed surgical oncology ICU	A daily-goals form completed as part of daily rounds was implemented. All providers, physicians, nurses, respiratory	Understanding of daily goals for care by nurses and residents; ICU length of stay.	The number of residents and nurses reporting that they understood the daily goals (4 or 5 on a 5-point scale) for a random sample of patients increased from less than 10% at baseline to over 95% by eight weeks after implementation. Over the course of a year, average ICU length of stay

Table A.2—Continued

Author	Year	Population	Intervention	Outcome Measures	Findings
			therapists, and pharmacists initialed the form 3 times per day. Form was updated if goals for the day changed.		decreased from 2.2 days to 1.1 days. No statistical tests were reported.
Pronovost et al.	2006	103 Michigan hospital ICUs in 67 hospitals reporting data for 1,981 ICU-months and 375,757 catheter-days	This was a multifaceted intervention, one component of which was a daily goal sheet to improve clinician-to-clinician communication within the ICU. ICUs also designated at least one physician and one nurse as team leaders.	Rates of catheter-related bloodstream infections	The multilevel Poisson regression model showed a significant decrease in rates of infection during all study periods compared with baseline rates. Incidence-rate ratios decreased continuously from 0.62 at 0 to 3 months to 0.34 at 16 to 18 months. The median rate of infection decreased from 2.7 per 1,000 catheter-days at baseline to 0 within the first three months after implementation ($p=.002$). The mean rate per 1,000 catheter-days decreased from 7.7 at baseline to 1.4 at 16 to 19 months of follow-up ($p<.002$).
Uhlig et al.	2002	Cardiac Surgery Program at one hospital	The implementation of a daily team-based collaborative rounds that uses a structured communication protocol and involves patients and families.	Unadjusted mortality, patient satisfaction, physician rating of work-life quality	Mortality was less than half the expected rates derived from pre-collaborative rounds experience (2.1% versus 4.8%), not tested statistically. On a 5-point scale, cardiac staff who participated in both collaborative and traditional rounds rated collaborative rounds more favorably on sense of common purpose (4.6 versus 3.9), collective power (4.0 versus 3.4), communication (4.5 versus 3.9), shared leadership responsibility (4.1 versus 3.4), problem solving (4.3 versus 3.3), respect (4.1 versus 3.4); and sense of collaboration (4.5 versus 3.4). Patient satisfaction with cardiac surgery care was rated in the 97th–99th percentile nationally. Statistical tests and pre-collaborative rounds satisfaction were not reported.
Young et al.	1998	469 ventilator-dependent patients	Collaborative daily bedside rounds; weekly	Risk-adjusted (APACHE II) ICU	There were reductions in risk-adjusted (APACHE II) ICU length of stay (19.8 versus 14.7 days,

Table A.2—Continued

Author	Year	Population	Intervention	Outcome Measures	Findings
		in a 12-bed medical-surgical ICU in a nonteaching tertiary care hospital, including 95 consecutive pre-intervention patients and 374 consecutive post-intervention patients	meetings with patient's family; creation of a multidisciplinary group to develop plans to standardize care processes, introduce interdisciplinary guidelines and protocols, and standardize order sheets.	length of stay, risk-adjusted mortality, hospital length of stay, hospital charges and costs.	p=.001), hospital length of stay (34.6 versus 25.9 days, p=.001), hospital charges ($102,500 versus $78,500, p=.001) and costs ($71,900 versus. $58,000, p=.001). There was not a significant change in risk-adjusted mortality.

65

Table A.3. Summary of Team-Training Studies That Examined Behaviors or Organizational Impacts

Studies not evaluating organizational impacts

Author	Year	Population	Study Design	Teamwork-Training Content	Simulators Used in Evaluation?	Findings
Blum et al.	2004	148 anesthesia faculty from four hospitals affiliated with medical school	Quasi-experimental: participants rated the course's acceptance, utility, and the need for recurrent training immediately after the course and approximately 1 year later. Participants also rated self-perceived changes in ability to manage difficult cases.	A scenario followed by an airline video displaying Crisis Resource Management (CRM) principles, and a didactic lecture focusing on role clarity, communication, personnel support, managing resources, and global assessment. This was followed by review of a videotape of initial scenario and a debriefing, in addition to participation in up to three additional scenarios with debriefings, one of which included a medical error	Yes	On a 5-point scale, participants rated the course as being high quality (4.84) and realistic (4.43), and the lecture as being valuable (4.51). They reported it helped them perform better (4.57), debrief better (4.48), and communicate better (4.60). While still highly rated a year later by the 55 responding participants, there were statistically significant degradations in ratings of debriefing better (4.04, $p<.008$) and communicating better (4.36, $p<.035$). Respondents who experienced a difficult event since the course reported the course improved many of their teamwork skills. Over 70% of participants reported that training should be repeated every 24 months or less.
Cashman et al.	2004	Interdisciplinary team in a community health center	Team members' attitudes were measured at baseline, one year, and two years using the SYMLOG (Systematic Multiple Level Observation of Groups) survey instrument	Five formal team-training workshops (team assessment, communication, problem solving, leadership) and increased team meeting time	No	In the first year of training, the SYMLOG assessment detected a positive change in teamwork behaviors, resulting in members feeling stronger bonds, greater commitment to helping one another, and a deeper understanding of team development. The second-year measurement showed some slippage in team development.

Table A.3—Continued

Author	Year	Population	Study Design	Teamwork-Training Content	Simulators Used in Evaluation?	Findings
Hope et al.	2005	34 participants of program, discipline not specified; 23 follow-up interviews.	Quasi-experimental: matched pre- and post-intervention evaluations completed by students. The Downstate Team Building Initiative is a year-long, 18-session curriculum for a multidisciplinary group of students from State University of New York–SUNY Downstate's medical, nursing, and allied health schools; 11 sessions combine didactic and interactive activities; during the remaining 7, students identify and implement a community health project. Also qualitative follow-up with students after they are in clinical practice.	Group decisionmaking, conflict mediation, and alliance building	No	On a 7-point scale, participants rated their post-intervention team-building skills more highly than their pre-intervention skills. Team atmosphere improved 48% ($p<.001$), group teamwork skills improved 44% ($p<.001$), individual teamwork skills improved 35% ($p<.001$), group multicultural skills improved 36% ($p<.001$), individual multicultural skills improved 13% ($p<.002$), understanding of interdisciplinary professional function improved 36% ($p<.001$); physician assistants and medical students had the greatest improvements. Understanding of professional training improved 52% ($p<.001$).
Howard et al.	1992	38 American anesthesiology	Quasi-experimental: participants were	Dynamic decisionmaking, planning leadership,	Yes	On a 5-point scale, participants reported that they enjoyed the

67

Table A.3—Continued

Author	Year	Population	Study Design	Teamwork-Training Content	Simulators Used in Evaluation?	Findings
		residents and faculty	trained, then their performance was observed during simulation; a debriefing session followed. Participants completed pre- and post-session surveys and tests on crisis-management principles and management of perioperative critical incidents, in addition to a 2-month post-training survey on the value of the course in their practice.	communication, workload distribution. Training consisted of half a day of lectures and group discussions and a 2 hour stimulator session		course (5.0) and reported that it was intense (4.0) and helpful to their practice (5.0). Residents improved in their knowledge about anesthesia crisis management; faculty did not (possible ceiling effect due to high level of knowledge before the training). During the simulations, poor communication, lack of task coordination, and lack of task delegation by leader was observed. However, there was not a pre-training baseline to determine whether the frequency of poor teamwork behaviors had declined.
Sica et al.	1999	24 radiology residents	Half of the residents participated in simulations of crisis scenarios; the other half observed. All residents then attended a lecture and watched a videotaped review. The residents then switched roles for another simulation. Simulation videotapes were reviewed and	Communication, use of resources and personnel, role clarity	Yes	Participants rated the course favorably on overall usefulness, attainment of course goals, realism of scenarios, quality of lecture, and quality of videotape review. Participants who attended lecture and watched videotape prior to participating in scenario performed better than the other residents in communication skills, use of support personnel, use of resources, role clarity, and global assessment.

Table A.3—Continued

Author	Year	Population	Study Design	Teamwork-Training Content	Simulators Used in Evaluation?	Findings
			residents rated on behavioral performance. Residents rated the course.			
Studies evaluating organizational impacts						
Awad et al.	2005	An unreported number of OR nurses, anesthesiologists and surgeons at one institution	Quasi-experimental: participants rated communication; use of preoperative briefings, appropriate usage and timing of administration of prophylactic antibiotics; deep venous thrombosis prophylaxis also assessed.	Didactic instruction, interactive participation, role-play, training films, and clinical vignettes. Specific topics not specified. Training performed by VA's National Center for Patient Safety. A change team that met weekly was also implemented.	No	Self-rated communication improved among anesthesiologists (p=.0008) and surgeons (p=.0004), but not OR nurses. Use of preoperative briefings increased from 64% at 1-month post-implementation to 100% by 4 months post-implementation. There were significant increases in the percentage of patients receiving prophylactic antibiotics within 60 minutes of incision and the number of patients receiving deep venous thrombosis prophylaxis before induction (p-values not reported).
Barrett et al. Morey et al.	2001 2002	684 American physicians, nurses and technicians from emergency departments in 6 hospitals. 374 physicians, nurses and technicians from emergency departments in	Quasi-experimental with control group. One pre-intervention and 2 post-intervention observation periods. Staff in all 9 EDs completed pre- and post-intervention surveys on attitudes toward teamwork and perception of support.	Maintenance of team structure and climate, planning and problem solving, conflict resolution, decisionmaking, shared mental models, communication, workload management, and improvement of tool skills (MedTeams®). Examples of skills taught include checkbacks and cross monitoring. It was an 8-hour course and practicum.	No	The quality of team behaviors improved, staff attitudes toward teamwork improved, and perceptions of support improved. Clinical errors significantly decreased. No significant improvements were found in subjective workload, preparation of patients for admission, or patient satisfaction. Only observed team behaviors were significantly improved relative to those of the

Table A.3—Continued

Author	Year	Population	Study Design	Teamwork-Training Content	Simulators Used in Evaluation?	Findings
		three hospitals acted as controls	Trained nurses and physicians rated teams on teamwork behaviors and clinical errors during the delivery of care. Admitting-unit nurses rated the preparation of ED patients for admission. Patient satisfaction with care delivered in ED was also measured.			control group.
DeVita et al.	2005	138 individuals trained, including 69 nurses, 21 respiratory therapists, and 48 physicians	Quasi-experimental: all teams trained, then their performance was observed and rated 3 times Videotape observation of behavior and technical performance ("survival" of a computerized patient, team, and facilitator rating of crisis task completion rate)	Web-based presentation; didactic review of key concepts of team performance (responding team members must be self-identified, tasks need to be characterized and prioritized, steps must be sequenced, roles and responsibilities need to be delegated and rehearsed); 3 simulation scenarios; and facilitator-moderated debriefings	Yes: advanced cardiac life support	Simulator "survival" improved from 0% to 90% across the 3 sessions in a day's course. Most of the improvement came between the first and second scenarios. The mean task-completion rate improved overall from 31% to 89%. Improvements were observed for each of the teams.
Dienst & Byl	1981	Phase 1: 52 medical, 22 nursing, 38 pharmacy, 2	Quasi-experimental: students participated in weekly student team-development	Review of history and origin of health care teams, communication, leadership, decisionmaking, role	No	Students improved in their knowledge of teamwork principles and patient problem-solving skills. There were not changes in their

Table A.3—Continued

Author	Year	Population	Study Design	Teamwork-Training Content	Simulators Used in Evaluation?	Findings
		dental, 2 physical therapy students. Phase 2: 110 medical, 35 nursing, 74 pharmacy, 7 other allied health students	seminars and participated in 2- or 3-person student teams. Phase 1 evaluation included assessments of students' attitudes toward team approach in health care, knowledge of teamwork principles, ability to apply team concepts to hypothetical patients, and students' preference for team approach. Phase 2 evaluation included patient-satisfaction, service volume, and comprehensiveness of care.	relationship, negotiation, and practical aspects of patient care and individual team process		attitudes about or preferences for a team approach. Students saw a greater number of patients than during previous clerkships and were rated as providing more-comprehensive care, based on chart reviews.
Gaba et al.	1998 and 2001	68 American anesthesiology residents and attendings	Quasi-experimental: participants received didactic training and participated in a simulation. A debriefing session followed. Participants completed a survey. Observers rated technical performance and	Leadership, communication, workload distribution, decisionmaking, and human error	Yes: cardiac arrest and malignant hyper-thermia scenarios	Both attendings and residents rated course highly, responding that they "learned a lot," felt that it could help them practice more safely, would be a good mechanism for improving team coordination, and felt it would be beneficial to other anesthesiologists. Self-rated performance in the simulations increased with increasing years of experience. The debriefing session

Table A.3—Continued

Author	Year	Population	Study Design	Teamwork-Training Content	Simulators Used in Evaluation?	Findings
			behavioral performance from videotapes of the simulations.			was highly rated, but the didactic sessions were less well received. Technical performance for all teams was rated fairly high, with moderate variability between teams. There was substantial variability of the behavioral performance, with a substantial number of teams and primary anesthesiologists being rated at the level of "minimally acceptable" or lower. While statistical measures of correlation between technical and behavioral performance were not presented in the article, graphed scores suggest a positive relationship.
i Gardi et al.	2001	32 Danish teams composed of one anesthesiologist and one nurse anesthetist	Quasi-experimental: all teams trained, then their performance was observed and rated. Videotaped observation of behaviors and technical performance	Leadership, communication, and other human-performance factors	Yes: malignant hyperthermia during anesthesia	All teams were able to diagnose malignant hyperthermia during the scenario. Although all teams intended to hyperventilate the patient, only 14 teams did so successfully (44%). Problems teams experienced were related to poor management of resources rather than lack of knowledge.
Jacobsen et al.	2001	42 Scandinavian anesthetists in two-person teams (21 teams)	Quasi-experimental: all teams trained, then their performance was observed and rated.	Awareness, leadership, communication, delegation, cooperation, declaration, re-evaluation, allocation, help hands, help competence, start of initial treatment	Yes: anaphylactic reaction during anesthesia	Teams were unable to make correct diagnosis without prompting from instructor. While teams demonstrated communication, little leadership or delegation was demonstrated.

Table A.3—Continued

Author	Year	Population	Study Design	Teamwork-Training Content	Simulators Used in Evaluation?	Findings
			Videotaped observation of behavior and technical performance			
Rivers, Swain, and Nixon	2003	164 surgical staff members at Methodist University Hospital	Quasi-experimental: participants completed a survey after the training. The number of surgical count errors for the six months before training was compared to the number for the six months after training.	There were 3 4-hour phases, each of which included multiple case studies, interactive team activities and skill applications, video examples of best performance, and knowledge testing. Topics included were effective team building, recognition of adverse situations, conflict management, stress management, decision-making, performance feedback, cross check and challenge, and fatigue management.	No	More than 75% of participants strongly or very strongly agreed that the training provided useful knowledge for improving job skills. 81% strongly or very strongly agreed that training would increase their effectiveness in the OR. The number of surgical count errors decreased by 50%; the actual number of surgical count errors was not reported nor was a statistical test presented.

Appendix B.
Members of the Clinical Advisory Panel

Paul Barach, MD, MPH
Until June 2008:
Visiting Professor
Anesthesia and Emergency Medicine
Utrecht University Medical Center
NETHERLANDS

Associate Professor
Departments of Anesthesiology and Epidemiology
University of South Florida College of Medicine

Vincent Carr, DO, FACC, FACP
Colonel, USAF, MC, CFS
Assistant Professor of Medicine
Uniformed Services University of Health Sciences

Patricia Collins, RN, MSN
Program Manager Family-Centered Care
Office of the Chief Medical Officer
TRICARE Management Activity

Paul Gluck, MD
Chair of the Board
National Patient Safety Foundation
Associate Clinical Professor
University of Miami Miller School of Medicine

Marlene Goldman, ScD
Professor, Obstetrics & Gynecology
and Community & Family Medicine
Dartmouth Medical School
Director of Clinical Research, OB/GYN
Dartmouth-Hitchcock Medical Center

Darryl T. Gray, MD, ScD
Medical Officer
Center for Quality Improvement and Patient Safety
Agency for Healthcare Research and Quality

Shukri Khuri, MD
Professor of Surgery
Harvard Medical School
VA Boston Healthcare System

Susan Mann, MD
Director of Quality Improvement–Ob/Gyn Department
Beth Israel Deaconess Medical Center
Assistant Clinical Professor, Harvard Medical School

Susan Meikle, MD, MSPH
Medical Officer
CRHB/NICHD/NIH

Pamela H. Mitchell, RN, PhD, FAAN, FAHA
Elizabeth S. Soule Professor & Associate Dean for Research, School of Nursing
Director, Center for Health Sciences Interprofessional Education
University of Washington

Peter E. Nielsen, MD, FACOG, COL, MC, USA
Chief, Department of Obstetrics and Gynecology
OB/GYN Consultant to The Surgeon General
Madigan Army Medical Center

Mary L. Salisbury RN, MSN
President
The Cedar Institute, Inc.

J. Bryan Sexton, PhD
Assistant Professor in Anesthesiology and Critical Care Medicine
The Johns Hopkins University School of Medicine.

Robert L. Wears, MD, MS
Professor, Department of Emergency Medicine
University of Florida

John S. Webster, MD, MBA
CAPT, Medical Corps, USN (Ret.)
President, Healthcare Consulting, Inc.

Appendix C.
Highly Rated Measures for Each Clinical Area

This appendix lists the measures that the advisory panel identified as being highly related to teamwork for each of the three clinical areas. The list is separated into groups of process measures and outcome measures. Within each group, we identify measures that were

- successfully tested with the DoD administrative health care data
- unable to be tested with the DoD health care data due to lack of required information; the lack was discovered during the testing process
- identified in advance as requiring information not contained in the DoD health care data, and therefore not tested using these data.

Highly Rated Measures for Labor and Delivery

Process Measures

Measures successfully tested:

No process measures capable of being used with encounter data.

Measures unable to be tested due to lack of detailed pharmacy data during inpatient stay:

Receipt of Group B Streptococcus (GBS) Antibiotics (Clinical Advisory Panel)

Women who give birth by cesarean section should receive at least one dose of antibiotic prophylaxis (McGlynn et al., 2000)

Highly rated measures, but not feasible with administrative data:

Time Elapsed from Decision to Incision for Stat C-Section (Risser et al., 2002)

Time Elapsed from Decision to Incision for Urgent C-Section (Risser et al., 2002)

Time Elapsed from Registration to Maternal-Fetal Assessment (Risser et al., 2002)

Time Elapsed from Scheduled Appointment to Actual Start Time of C-Section (Risser et al., 2002)

Time Elapsed from Request to Administration of Regional Anesthesia (Risser et al., 2002)

Outcome Measures

Measures successfully tested:

Birth Trauma, Injury to Neonate–C-Sections (Subset of PSI 17) (AHRQ, 2007b)

Birth Trauma, Injury to Neonate–Vaginal Birth (Subset of PSI 17) (AHRQ, 2007b)

Uterine Rupture (Mann et al., 2006)

Maternal Death (Mann et al., 2006)

Intrapartum Fetal Death (term baby) (Mann et al., 2006)

Measures unable to be tested due to inability to distinguish between planned and unplanned C-Section:

Frequency of General Anesthesia in an Unplanned C-Section (Mann et al., 2006)

Highly rated measures, but not feasible with administrative data:

Unplanned Admission to Intensive Care (Mann et al., 2006)

Hospital-Level Procedure Utilization Rates for Primary C-Section Delivery (Inpatient Quality Indicator [IQI] 33) (AHRQ, 2007a)

Unplanned Return to Labor and Delivery Unit or Operating Room (Mann et al., 2006)

5-Minute Apgar Score < 7 (term baby) (Mann et al., 2006)

Highly Rated Measures for Surgery

Process Measures

Measures successfully tested:

No process measures capable of being constructed with encounter data.

Highly rated measures, but not feasible with administrative data:

Patients having surgical repair of a hip fracture offered a complete medical evaluation preoperatively, including (a) medical history, (b) physical examination, (c) laboratory evaluation, (d) electrocardiogram (RAND)—(Kerr et al., 2000a)

Percentage of Major Cardiac Surgical Patients with Controlled Perioperative Serum Glucose (Surgical Care Improvement Project—SCIP, 2007)

Anti-Platelet Medications at Discharge–Cardiac Surgery (Society of Thoracic Surgeons [STS]) (National Quality Forum, 2007)

Prophylactic Antibiotic Received 1 Hour Prior to Incision–Overall Rate (Joint Commission, 2007a)

Prophylactic Antibiotic Received 1 Hour Prior to Incision–CABG (Joint Commission, 2007a)

Prophylactic Antibiotic Received 1 Hour Prior to Incision–Cardiac Surgery (Joint Commission, 2007a)

Prophylactic Antibiotics Received 1 Hour Prior to Incision–Colon Surgery (Joint Commission, 2007a)

Prophylactic Antibiotics Received 1 Hour Prior to Incision–Hysterectomy (Joint Commission, 2007a)

Prophylactic Antibiotics Received 1 Hour Prior to Incision–Vascular Surgery (Joint Commission, 2007a)

Preoperative Beta Blockade for Cardiac Surgery (STS) (National Quality Forum, 2007)

Percentage of Patients with Known Coronary Artery Disease or Other Atherosclerotic Cardiovascular Disease Diagnoses, Without Contraindications to Beta Blockers, Who Received Beta Blockers During the Perioperative Period (SCIP, 2007)

Percentage of Major Surgery Patients Maintained on a Beta Blocker Prior to Surgery Who Received a Beta Blocker During the Perioperative Period (SCIP, 2007)

Adequate Post-Discharge Plan for Patients with CABG—Follow-up Visit and Prescription for Aspirin or Persantine Unless on Coumadin (VA, 2007)

Prophylactic Antibiotics Discontinued Within 24 Hours After Surgery End Time–CABG (Joint Commission, 2007a)

Prophylactic Antibiotics Discontinued Within 24 Hours After Surgery End Time–Cardiac Surgery (Joint Commission, 2007a)

Prophylactic Antibiotics Discontinued Within 24 Hours After Surgery End Time–Hip Arthroplasty (Joint Commission, 2007a)

Prophylactic Antibiotics Discontinued Within 24 Hours After Surgery End Time–Knee Arthroplasty (Joint Commission, 2007a)

Prophylactic Antibiotics Discontinued Within 24 Hours After Surgery End Time–Colon Surgery (Joint Commission, 2007a)

Prophylactic Antibiotics Discontinued Within 24 Hours After Surgery End Time–Hysterectomy (Joint Commission, 2007a)

Prophylactic Antibiotics Discontinued Within 24 Hours After Surgery End Time–Vascular Surgery (Joint Commission, 2007a)

Outcome Measures

Measures successfully tested:

Failure to Rescue (PSI) (AHRQ, 2007b)

Foreign Body Left in During Procedure (AHRQ, 2007b)

Mortality in Low-Mortality DRG (AHRQ, 2007b)

Highly rated measures, but not feasible with administrative data:

Peripheral Nerve Injury (National Surgical Quality Improvement Program- [NSQIP]) (American College of Surgeons, 2007)

Unplanned Intubation (NSQIP) (American College of Surgeons, 2007)

Acute Renal Failure (NSQIP) (American College of Surgeons, 2007)

DVT Thrombophlebitis (NSQIP) (American College of Surgeons, 2007)

Average Hospital Length of Stay (NSQIP) (American College of Surgeons, 2007)

Average Surgical Length of Stay (NSQIP) (American College of Surgeons, 2007)

Risk-Adjusted CABG In-Patient Mortality (STS) (NQF, 2007)

Pulmonary Embolism (NSQIP) (American College of Surgeons, 2007)

On Ventilator >48 Hours (NSQIP) (American College of Surgeons, 2007)

Septic Shock (NSQIP) (American College of Surgeons, 2007)

Return to the OR Within 30 Days (NSQIP) (American College of Surgeons, 2007)

Death Within 30 Days of Procedure (NSQIP) (American College of Surgeons, 2007)

Deep Incisional Surgical-Site Infection (SSI) (NSQIP) (American College of Surgeons, 2007)

Pneumonia (NSQIP) (American College of Surgeons, 2007)

Cerebrovascular Accident (CVA) Stroke (NSQIP) (American College of Surgeons, 2007)

Cardiac Arrest Requiring Cardiopulmonary Resuscitation (CPR) (NSQIP) (American College of Surgeons, 2007)

Myocardial Infarction (NSQIP) (American College of Surgeons, 2007)

Bleeding >4 Units Packed Red Blood Cells (PRBCs) or Whole Blood Transfusions (NSQIP) (American College of Surgeons, 2007)

Graft/Prosthesis/Flap Failure (NSQIP) (American College of Surgeons, 2007)

Systemic Inflammatory Response Syndrome (SIRS) (NSQIP) (American College of Surgeons, 2007)

HIGHLY RATED MEASURES FOR ACUTE MYOCARDIAL INFARCTION

Process Measures

Measures unable to be tested due to lack of detailed pharmacy data during inpatient stay:

Patients admitted with AMI should not receive short-acting nifedipine during the hospitalization (RAND)—(Kerr et al., 2000b)

Patients admitted with AMI should not receive any calcium channel blocker if they have a reduced left ventricular ejection fraction (LVEF) (≤40%) or heart failure during the hospitalization (RAND) (Kerr et al., 2000b)

Patients discharged after an AMI who have an LVEF≤40% documented during hospitalization should receive ACE inhibitors at discharge (unless they have contraindications to ACE inhibitors) (RAND)—(Kerr et al., 2000b)

Measures unable to be tested due to lack of detailed laboratory data during inpatient stay:

Cholesterol Assessment During Hospital Stay (CMS Test) (Joint Commission, 2007a)

Highly rated measures, but not feasible with administrative data:

At least 160 mg of aspirin within 2 Hours of presentation or admission unless patient has contraindications (RAND)—(Kerr et al., 2000b)

Aspirin prescribed at discharge at a dose of at least 81mg/day (RAND)—(Kerr et al., 2000b)

Mean Time from Arrival to Thrombolysis (Joint Commission, 2007a)

Thrombolytic Agent Received within 30 Minutes of Hospital Arrival (Joint Commission, 2007a)

ACE Inhibitor or Angiotension II Receptor Blocker (ARB) for Left Ventricular Systolic Dysfunction (LVSD) at Discharge (Joint Commission, 2007a)

Percentage of patients hospitalized with acute coronary syndrome found to be ST-segment elevation MI patients who met criteria for reperfusion and received interventional reperfusion (VA, 2007)

Aspirin Prescribed at Discharge (Joint Commission, 2007a)

Beta Blocker Prescribed at Discharge (Joint Commission, 2007a)

Patients <75 years old presenting with AMI who are within 12 hours of onset of symptoms and who do not have contraindications to thrombolysis or revascularization should receive a thrombolytic within 1 hour of the time their electrocardiogram (ECG) initially shows either of the following: (a) ST elevation >0.1 mV in 2 or more contiguous leads or (b) a left bundle branch block (LBBB) not known to be old (RAND) (Kerr EA et al., 2000b)

Adult Smoking Cessation Advice/Counseling (Joint Commission, 2007a)

Patients hospitalized for MI or to rule out MI should have a 12-lead ECG within 20 minutes of presentation (RAND)—(Kerr et al., 2000b)

Percentage of patients with first troponin result returned within 60 minutes of order time (VA, 2007)

Beta Blocker within 12 Hours of Admission (RAND)—(Kerr et al., 2000b)

Percentage of patients hospitalized with acute coronary syndrome found to be low-to-moderate-risk ACS patients who received a non-invasive stress test prior to discharge (VA, 2007)

Percentage of patients hospitalized with acute coronary syndrome (ACS) found to be moderate-to-low-risk ACS patients who had a plan prior to discharge that includes further outpatient stress testing and possible catheterization (VA, 2007)

Mean Time from Arrival to Percutaneous Coronary Intervention (PCI) (Joint Commission, 2007a)

PCI Received within 120 Minutes of Hospital Arrival (Joint Commission, 2007a; VA, 2007)

Patients admitted within 12 hours of the onset of AMI who do not have contraindications to heparin should receive heparin for at least 24 hours, unless they have received streptokinase, anisoylated plasminogen streptokinase activator complex (APSAC), or urokinase—(RAND) (Kerr et al., 2000b)

Lipid-Lowering Therapy Prescribed at Discharge for Patients with Elevated Low-Density Lipoprotein (LDL) Cholesterol (CMS Test) (Joint Commission, 2007a)

LDL Cholesterol Testing within 24 Hours After Hospital Arrival (CMS Test) (Joint Commission, 2007a)

Beta Blocker Within 24 Hours of Arrival (Joint Commission, 2007a)

Aspirin Within 24 Hours Before or After Hospital Arrival (Joint Commission, 2007a)

Percentage of patients hospitalized with acute coronary syndrome with cardiac symptoms prior to or on arrival to the acute care setting who had an ECG performed within 10 minutes after arrival (VA, 2007)

Outcome Measures

No outcome measures were highly rated by the clinical advisory panel.

Appendix D.
Descriptions of Highly Rated Measures for Each Clinical Area

Table D.1. Highly Rated Measures for Labor and Delivery

Measure	Source	Numerator	Denominator
Process Measures			
Receipt of GBS Antibiotics	Clinical Advisory Panel	Number of GBS-positive women or women at risk for GBS who are given antibiotics during labor	Women who are GBS-positive or at risk for GBS
Antibiotic Prophylaxis for C-Section	McGlynn et al., 2000	Number of women giving birth by C-section who receive at least one dose of antibiotic prophylaxis	Number of women giving birth by C-section at the facility
Time Elapsed from Decision to Incision for Stat C-Section	Risser et al., 2002	Sum of time elapsed from decision to incision for stat C-section patients	Number of stat C-section patients
Time Elapsed from Decision to Incision for Urgent C-Section	Risser et al., 2002	Sum of time elapsed from decision to incision for urgent C-section patients	Number of urgent C-section patients
Time Elapsed from Registration to Maternal-Fetal Assessment	Risser et al., 2002	Sum of time elapsed from registration to maternal-fetal assessment for women moving from registration to maternal-fetal assessment	Number of women moving from registration to maternal-fetal assessment
Time Elapsed from Scheduled Appointment to Actual Start Time of C-Section	Risser et al., 2002	Sum of time elapsed from scheduled appointment to actual start of C-section for women having scheduled C-sections	Number of women with scheduled C-sections who have C-sections
Time Elapsed from Request to Administration of Regional Anesthesia	Risser et al., 2002	Sum of time elapsed from request to administration of regional anesthesia for women requesting and receiving regional anesthesia	Number of women who request and receive regional anesthesia

Table D.1—Continued

Measure	Source	Numerator	Denominator
Outcome Measures			
Birth Trauma, Injury to Neonate—Cesarean Section Subset of PSI 17	AHRQ, 2007b	Number of discharges in denominator with ICD-9-CM code for birth trauma in any diagnosis field. **Exclude:** infants with a subdural or cerebral hemorrhage (subgroup of birth-trauma coding) any diagnosis code of preterm infant (denoting birth weight of less than 2,500 grams and less than 37 weeks' gestation or 34 weeks' gestation or less). infants with injury to skeleton and any diagnosis code of osteogenesis imperfecta.	Number of infants delivered via cesarean section
Birth Trauma, Injury to Neonate—Vaginal Delivery Subset of PSI 17	AHRQ, 2007b	Number of discharges in denominator with ICD-9-CM code for birth trauma in any diagnosis field. **Exclude:** infants with a subdural or cerebral hemorrhage (subgroup of birth-trauma coding) any diagnosis code of preterm infant (denoting birth weight of less than 2,500 grams and less than 37 weeks' gestation or 34 weeks' gestation or less). infants with injury to skeleton and any diagnosis code of osteogenesis imperfecta.	Number of infants delivered vaginally
Uterine Rupture	Mann et al., 2006	Number of women who experience uterine rupture during delivery	Number of women who delivered infants
Maternal Death	Mann et al., 2006	Number of women who expire as a result of giving birth	Number of women giving birth at the facility
Intrapartum Fetal Death (term baby)	Mann et al., 2006	Number of intrapartum fetal deaths	Number of births at the facility
Frequency of General Anesthesia in an Unplanned C-Section	Mann et al., 2006	Number of women given general anesthesia in an unplanned cesarean section	Number of women receiving unplanned cesarean sections

Table D.1—Continued

Measure	Source	Numerator	Denominator
Unplanned Admission to Intensive Care	Mann et al., 2006	Number of women admitted to an intensive care unit as a result of giving birth	Number of women giving birth at the facility
Hospital-Level Procedure Utilization Rates for Primary C-Section Delivery (IQI 33)	AHRQ, 2007a	Number of cesarean deliveries, identified by DRG, or by ICD-9-CM procedure codes if they are reported without a hysterotomy procedure.	All deliveries. **Exclude:** • patients with abnormal presentation • preterm delivery • fetal death • multiple gestation diagnosis codes • breech procedure codes • a previous cesarean-delivery diagnosis in any diagnosis field.
Unplanned Return to Labor and Delivery Unit or Operating Room	Mann et al., 2006	Number of women returned to the labor and delivery unit or the operating room, on an unplanned basis, after giving birth	Number of women giving birth at the facility
5-Minute Apgar Score < 7 (term baby)	Mann et al., 2006	Number of 5-minute newborn Apgar scores of less than 7 for term babies	Number of births at the facility

Table D.2. Highly Rated Measures for Surgery

Measure	Source	Numerator	Denominator
Process Measures			
Complete medical evaluation preoperatively for repair of hip fracture	Kerr et al., 2000a	Patients having surgical repair of a hip fracture who are given a medical history, physical exam, laboratory evaluation, and electrocardiogram	All patients undergoing surgical repair of a hip fracture
Controlled Perioperative Serum Glucose–Cardiac Surgery	SCIP, 2007	Surgery patients with controlled 6 a.m. serum glucose (≤ 200 mg/dL)	Cardiac surgery patients with no evidence of prior infection **Exclude:** • Patients who had a principal diagnosis suggestive of preoperative infectious diseases • Patients less than 18 years of age • Patients with physician-documented infection prior to surgical procedure of interest • Burn and transplant patients • Patients whose ICD-9-CM principal procedure occurred prior to the date of admission
Anti-platelet Medications at Discharge–Cardiac Surgery (STS)	NQF, 2007	Number of patients who were discharged on aspirin/safety-coated aspirin or clopidogrel after isolated CABG	All patients undergoing isolated CABG
Prophylactic Antibiotic Received 1 Hour Prior to Incision–Overall Rate	Joint Commission (JC) 2007a	Number of surgical patients who received prophylactic antibiotics within one hour before surgical incision (two hours if receiving vancomycin or a fluoroquinolone)	All selected surgical patients with no evidence of prior infection **Exclude:** • Patients who had a principal or admission diagnosis suggestive of preoperative infectious diseases • Patients who were receiving antibiotics at the time of admission (except colon surgery patients taking oral prophylactic antibiotics) • Patients who were receiving antibiotics more than 24 hours prior to surgery (except colon surgery patients taking oral prophylactic antibiotics • Colon surgery patients who received oral

Table D.2—Cont.

Measure	Source	Numerator	Denominator
			prophylactic antibiotics only, and who received no antibiotics during stay • Patients who are less than 18 years of age • Patients with physician-documented infection prior to surgical procedure of interest
Prophylactic Antibiotic Received 1 Hour Prior to Incision—CABG	JC, 2007a	Number of CABG patients who received prophylactic antibiotics within one hour before surgical incision (two hours if receiving vancomycin or a fluoroquinolone)	All CABG patients with no evidence of prior infection
Prophylactic Antibiotic Received 1 Hour Prior to Incision—Cardiac Surgery	JC, 2007a	Number of cardiac surgery patients who received prophylactic antibiotics within one hour before surgical incision (two hours if receiving vancomycin or a fluoroquinolone)	All cardiac surgery patients with no evidence of prior infection
Prophylactic Antibiotics Received 1 Hour Prior to Incision—Colon Surgery	JC, 2007a	Number of colon surgery patients who received prophylactic antibiotics within one hour before surgical incision (two hours if receiving vancomycin or a fluoroquinolone)	All colon surgery patients with no evidence of prior infection
Prophylactic Antibiotics Received 1 Hour Prior to Incision—Hysterectomy	JC, 2007a	Number of hysterectomy patients who received prophylactic antibiotics within one hour before surgical incision (two hours if receiving vancomycin or a fluoroquinolone)	All hysterectomy patients with no evidence of prior infection
Prophylactic Antibiotics Received 1 Hour Prior to Incision—Vascular Surgery	JC, 2007a	Number of vascular surgery patients who received prophylactic antibiotics within one hour before surgical incision (two hours if receiving vancomycin or a fluoroquinolone)	All selected vascular surgery patients with no evidence of prior infection
Pre-Operative Beta Blockade for Cardiac Surgery (STS)	NQF, 2007	Number of patients coming to isolated CABG with documented preoperative (24 hours) beta blockade	All patients undergoing isolated CABG

Table D.2—Cont.

Measure	Source	Numerator	Denominator
Patients with Known CAD or Other Atherosclerotic Cardiovascular Disease Diagnoses, Without Contraindications to Beta Blockers, Who Received Beta Blockers During the Perioperative Period	SCIP, 2007	Patients with known coronary artery disease (CAD) or other atherosclerotic cardiovascular disease diagnoses, without contraindications to beta blockers, who received beta blockers during the perioperative period	Surgery patients with known CAD or other atherosclerotic cardiovascular disease diagnoses
Surgery Patients on Beta Blocker Therapy Prior to Admission Who Received a Beta Blocker During the Perioperative Period	SCIP, 2007	Surgery patients on beta blocker therapy before admission who receive a beta blocker during the perioperative period	All surgery patients on beta blocker therapy prior to admission
Adequate Post-Discharge Plan for Patients with CABG	VA, 2007	Patients with evidence that a follow-up visit would be scheduled and with a prescription for aspirin or Persantine, unless on coumadin	All patients undergoing CABG
Prophylactic Antibiotics Discontinued Within 48 Hours After Surgery End Time—CABG	JC, 2007a	Number of CABG patients whose prophylactic antibiotics were discontinued within 48 hours after surgery end time	All CABG patients with no evidence of prior infection
Prophylactic Antibiotics Discontinued Within 48 Hours After Surgery End Time—Cardiac Surgery	JC, 2007a	Number of cardiac surgery patients whose prophylactic antibiotics were discontinued within 48 hours after surgery end time	All cardiac surgery patients with no evidence of prior infection
Prophylactic Antibiotics Discontinued Within 24 Hours After Surgery End Time—Hip Arthroplasty	JC, 2007a	Number of hip arthroplasty patients whose prophylactic antibiotics were discontinued within 24 hours after surgery end time	All hip arthroplasty patients with no evidence of prior infection
Prophylactic Antibiotics Discontinued Within 24 Hours After Surgery End Time—Knee Arthroplasty	JC, 2007a	Number of knee arthroplasty patients whose prophylactic antibiotics were discontinued within 24 hours after surgery end time	All knee arthroplasty patients with no evidence of prior infection
Prophylactic Antibiotics Discontinued Within 24 Hours	JC, 2007a	Number of colon surgery patients whose prophylactic antibiotics were	All colon surgery patients with no evidence of prior infection

Table D.2—Cont.

Measure	Source	Numerator	Denominator
After Surgery End Time—Colon Surgery		discontinued within 24 hours after surgery end time	
Prophylactic Antibiotics Discontinued Within 24 Hours After Surgery End Time—Hysterectomy	JC, 2007a	Number of hysterectomy patients whose prophylactic antibiotics were discontinued within 24 hours after surgery end time	All hysterectomy patients with no evidence of prior infection
Prophylactic Antibiotics Discontinued Within 24 Hours After Surgery End Time—Vascular Surgery	JC, 2007a	Number of vascular surgery patients whose prophylactic antibiotics were discontinued within 24 hours after surgery end time	All vascular surgery patients with no evidence of prior infection
Outcome Measures			
Failure to Rescue (PSI)[a]	AHRQ, 2007b	Discharges with a disposition of "deceased."	Discharges with potential complications of care listed in *failure to rescue* definition (i.e., pneumonia, DVT/PE, sepsis, acute renal failure, shock/cardiac arrest, or GI hemorrhage/acute ulcer). Exclusion criteria specific to each diagnosis. **Exclude:** • patients age 75 years and older. • neonatal patients in MDC 15. • patients transferred to an acute care facility. • patients transferred from an acute care facility. • patients admitted from a long-term care facility.
Foreign Body Left in During Procedure (PSI)	AHRQ, 2007b	Discharges with ICD-9-CM codes for foreign body left in during procedure in any secondary-diagnosis field.	All medical and surgical discharges defined by specific DRGs. **Exclude:** • Patients with ICD-9-CM codes for foreign body left in during procedure in the principal-diagnosis field
Mortality in Low-Mortality DRG (PSI)[b]	AHRQ, 2007b	Discharges with disposition of "deceased."	Patients in DRGs with less than 0.5% mortality rate. If a DRG is divided into "without/with complications," both DRGs must have mortality rates below 0.5% to qualify for inclusion. **Exclude:** • Patients with any code for trauma,

Table D.2—Cont.

Measure	Source	Numerator	Denominator
			immunocompromised state, or cancer.
Peripheral Nerve Injury (NSQIP)	ACS, 2007	Number of patients with peripheral nerve injury post surgery	Not determined (from available source documents)—none for event reporting; may be identified for definable populations
Unplanned Intubation (NSQIP)	ACS, 2007	Number of patients who undergo an unplanned intubation	Not determined (from available source documents)—none for event reporting; may be identified for definable populations
Acute Renal Failure (NSQIP)	ACS, 2007	Number of patients with acute renal failure who did not require dialysis preoperatively	Not determined (from available source documents)—none for event reporting; may be identified for definable populations
DVT Thrombophlebitis (NSQIP)	ACS, 2007	Number of patients with a DVT/Thrombophlebitis post surgery	Not determined (from available source documents)—none for event reporting; may be identified for definable populations
Average Hospital Length of Stay (NSQIP)	ACS, 2007	NA	
Average Surgical Length of Stay (NSQIP)	ACS, 2007	NA	
Risk-Adjusted CABG In-Patient Mortality (STS)	NQF, 2007	Number of patients undergoing isolated CABG who die, including both (1) all deaths during the hospitalization in which the operation was performed, even if after 30 days, and (2) those deaths occurring after discharge from the hospital, but within 30 days of the procedure, unless the cause of death is clearly unrelated to the operation	All patients undergoing isolated CABG
Pulmonary Embolism (NSQIP)	ACS, 2007	Number of patients with a pulmonary embolism post-surgery	Not determined (from available source documents)—none for event reporting; may be identified for definable populations
On Ventilator >48 hours (NSQIP)	ACS, 2007	Number of patients on a ventilator longer than 48 hours	Not determined (from available source documents)—none for event reporting; may be identified for definable populations
Septic Shock (NSQIP)	ACS, 2007	Number of patients with septic shock	Not determined (from available source documents)—none for event reporting; may be identified for definable populations

Table D.2—Cont.

Measure	Source	Numerator	Denominator
Return to the OR Within 30 Days (NSQIP)	ACS, 2007	Number of patients who return to the OR within 30 days	Not determined (from available source documents)—none for event reporting; may be identified for definable populations
Death Within 30 Days of Procedure (NSQIP)	ACS, 2007	Number of patients who die within 30 days of a procedure	Not determined (from available source documents)— none for event reporting; may be identified for definable populations
Deep Incisional SSI (NSQIP)	ACS, 2007	Number of patients with a deep incisional SSI	Not determined (from available source documents)—none for event reporting; may be identified for definable populations
Pneumonia (NSQIP)	ACS, 2007	Number of patients with non-preoperative pneumonia	Not determined (from available source documents)—none for event reporting; may be identified for definable populations
CVA Stroke (NSQIP)	ACS, 2007	Number of patients with a CVA/stroke post surgery	Not determined (from available source documents)—none for event reporting; may be identified for definable populations
Cardiac Arrest Requiring CPR	ACS, 2007	Number of patients with cardiac arrest requiring CPR post surgery	Not determined (from available source documents)—none for event reporting; may be identified for definable populations
Myocardial Infarction (NSQIP)	ACS, 2007	Number of patients with an acute MI during or after surgery	Not determined (from available source documents)—none for event reporting; may be identified for definable populations
Bleeding >4 units PRBCs or Whole Blood Transfusions (NSQIP)	ACS, 2007	Number of patients with bleeding requiring more than 4 units PRBCs within the first 72 hours postoperatively	Not determined (from available source documents)—none for event reporting; may be identified for definable populations
Graft/Prosthesis/Flap Failure (NSQIP)	ACS, 2007	Number of patients with graft/prosthesis/flap failure	Not determined (from available source documents)—none for event reporting; may be identified for definable populations
SIRS (NSQIP)	ACS, 2007	Number of patients with Systemic Inflammatory Response Syndrome	Not determined (from available source documents)—none for event reporting; may be identified for definable populations

[a] The specification for Failure to Rescue includes both surgical and medical patients. Empirical analyses for this project limited the eligible population to surgical patients.

[b] The specification for Mortality in Low Mortality DRG includes both surgical and medical patients. Empirical analyses for this project limited the eligible population for the measure to surgical patients.

NA = Not Available

Table D.3. Highly Rated Measures for Acute Myocardial Infarction

Measure	Source	Numerator	Denominator
Process Measures			
No Short-Acting Nifedipine	Kerr et al., 2000b	Patients who receive nifedipine when hospitalized with an AMI	Patients hospitalized with an AMI
No Calcium Channel Blocker if Reduced LVEF	Kerr et al., 2000b	Patients hospitalized with an AMI who have a reduced LVEF ($\leq$40%) or heart failure during the hospitalization and receive a calcium channel blocker	Patients hospitalized with an AMI
ACE Inhibitor at Discharge	Kerr et al., 2000b	Patients discharged after an AMI who have an LVEF$\leq$40% documented at any time during the hospitalization who receive ACE inhibitors at discharge unless they have contraindications to ACE inhibitors	Patients hospitalized with an AMI
Cholesterol Assessment During Hospital Stay	JC, 2007a	AMI patients with documentation of LDL-c level in the hospital record or documentation that LDL-c testing was done during the hospital stay or is planned after discharge	AMI patients **Exclude:** • Patients less than 18 years of age • Patients received in transfer from another acute care hospital • Patients who expired • Patients who left against medical advice • Patients discharged to hospice • Patients with reason documented for no LDL-c testing
At least 160mg of aspirin within 2 hours of presentation or admission, unless patient has contraindications	Kerr et al., 2000b	Patients who receive at least 160mg of aspirin within 2 hours of presentation	AMI patients **Exclude:** • Patients with contraindications to aspirin
Aspirin prescribed at discharge at a dose of at least 81mg/day	Kerr et al., 2000b	AMI patients who are prescribed aspirin at discharge	AMI patients without aspirin contraindications

Table D.3—Continued

Measure	Source	Numerator	Denominator
			Exclude: • Patients less than 18 years of age • Patients transferred to another acute care hospital or federal hospital on day of arrival • Patients discharged to hospice • Patients who expired • Patients who left against medical advice • Patients with one or more of the following aspirin contraindications/reasons for not prescribing aspirin documented in the medical record: active bleeding on arrival or within 24 hours of arrival; aspirin allergy; coumadin/warfarin prescribed at discharge; other reasons documented by a physician, nurse practitioner, or physician assistant for not prescribing aspirin at discharge
Mean Time from Arrival to Thrombolysis	JC, 2007a	Sum of time in minutes from hospital arrival to administration of thrombolytic agent in patients with ST segment elevation or LBBB on the ECG performed closest to hospital arrival time	AMI patients with ST-elevation or LBBB on ECG who received a thrombolytic agent **Exclude:** • Patients less than 18 years of age • Patients received in transfer from another acute care hospital, including another emergency department • Patients with "comfort measures only" documented by a physician, nurse practitioner, or physician assistant • Patients who did not receive a thrombolytic agent within 30 minutes and had a reason for delay documented by a physician, nurse practitioner, or physician assistant (e.g., social, religious, initial concern or refusal)
Thrombolytic Agent Received	JC,	AMI patients whose time from hospital arrival	AMI patients with ST-elevation or LBBB on

Table D.3—Continued

Measure	Source	Numerator	Denominator
Within 30 Minutes of Hospital Arrival	2007a	to administration of thrombolytic agent is 30 minutes or less	ECG who received a thrombolytic agent **Exclude:** • Patients less than 18 years of age • Patients received in transfer from another acute care hospital, including another emergency department • Patients with "comfort measures only" documented by a physician, nurse practitioner, or physician assistant • Patients who did not receive a thrombolytic agent within 30 minutes and had a reason for delay documented by a physician, nurse practitioner, or physician assistant (e.g., social, religious, initial concern or refusal)
ACE Inhibitor (ACEI) or ARB for left ventricular systolic disorder (LVSD) at Discharge	JC, 2007a	AMI patients who are prescribed an ACEI or ARB at hospital discharge	AMI patients with LVSD and without both ACEI and ARB contraindications **Exclude:** • Patients less than 18 years of age • Patients transferred to another acute care hospital or federal hospital on day of arrival • Patients discharged to hospice • Patients who expired • Patients who left against medical advice • Patients with chart documentation of participation in a clinical trial testing alternatives to ACEIs as first-line heart-failure therapy • Patients with *both* a potential contraindication/reason for not prescribing an ACEI at discharge *and* a potential contraindication/reason for not prescribing an ARB at discharge, as evidenced by one or more of the following: ACEI allergy *and* ARB allergy; moderate or severe aortic stenosis; documentation of *both* a reason for not

Table D.3—Continued

Measure	Source	Numerator	Denominator
			prescribing an ACEI at discharge *and* a reason for not prescribing an ARB at discharge; documentation for not prescribing an ARB at discharge *and* an ACEI allergy, documentation for not prescribing an ACEI at discharge *and* an ARB allergy
Patients hospitalized with acute coronary syndrome found to be ST-segment elevation MI (STEMI) who met criteria for reperfusion and received interventional reperfusion	VA, 2007	Patients hospitalized with ACS found to be STEMI who met criteria for reperfusion and received reperfusion	Patients hospitalized with ACS found to be STEMI who met criteria for reperfusion **Exclude:** • Patients transferred from a community hospital or ED • Documented decision not to treat within 24 hours. The record clearly documents that the patient, patient's family, or legal representative wishes comfort measures only, and/or there is agreement that the patient's cardiac condition and co-morbid conditions preclude aggressive treatment. Documentation-such as comfort measures only; hospice care; maintain treatment for comfort; terminal care; physician documentation that care is limited at family's request or due to patient's age or chronic illness, palliative care, supportive care only—will cause the patient to be excluded from the measure. • Contraindications to percutaneous transluminal coronary angioplasty (PTCA/PCI) – Includes documentation of patient or family refusal, decision not to treat, patient co-morbidities that preclude procedure, other reason documented
Aspirin Prescribed at Discharge	JC, 2007a	AMI patients who are prescribed aspirin at discharge	AMI patients without aspirin contraindications **Exclude:** • Patients less than 18 years of age • Patients transferred to another acute care hospital or federal hospital on day of arrival

97

Table D.3—Continued

Measure	Source	Numerator	Denominator
			• Patients discharged to hospice • Patients who expired • Patients who left against medical advice • Patients with one or more of the following aspirin contraindications/reasons for not prescribing aspirin documented in the medical record: active bleeding on arrival or within 24 hours of arrival; aspirin allergy; coumadin/warfarin prescribed at discharge; other reasons documented by a physician, nurse practitioner, or physician's assistant for not prescribing aspirin at discharge
Beta Blocker Prescribed at Discharge	JC, 2007a	AMI patients who are prescribed a beta blocker at discharge	AMI patients without beta blocker contraindications **Exclude:** • Patients less than 18 years of age • Patients transferred to another acute care hospital or federal hospital • Patients discharged to hospice • Patients who expired • Patients who left against medical advice • Patients with one or more of the following contraindications/reasons for not prescribing a beta blocker documented in the medical record: beta blocker allergy; bradycardia on day of discharge or day prior to discharge when not on beta blocker; second- or third-degree heart block on ECG on arrival or during hospital stay and does not have a pacemaker; systolic blood pressure less than 90mm Hg on day of discharge or day prior to discharge while not on beta blocker; other reasons documented by a physician, nurse practitioner or physician assistant for not prescribing a beta blocker

Table D.3—Continued

Measure	Source	Numerator	Denominator
Thrombolytic Within 1 Hour	Kerr et al., 2000b	Patients who meet the specified criteria and receive a thrombolytic within 1 hour of their ECG	Patients hospitalized with an MI or to rule out MI **Exclude:** • Patients with contraindications to thrombolysis or revascularization
Adult Smoking-Cessation Advice/Counseling	JC, 2007a	AMI patients (cigarette smokers) who receive smoking-cessation advice or counseling during the hospital stay	AMI patients with a history of smoking cigarettes anytime during the year prior to hospital arrival **Exclude:** • Patients less than 18 years of age • Patients transferred to another acute care hospital or federal hospital on day of arrival • Patients discharged to hospice • Patients who expired • Patients who left against medical advice
Patients hospitalized for MI or to rule out MI should have a 12-lead ECG within 20 minutes of presentation	Kerr et al., 2000b	Patients who receive a 12-lead ECG within 20 minutes of presentation	Patients hospitalized with an MI or to rule out MI
First troponin result returned within 60 minutes of order time	VA, 2007	Patients with result of troponin measurement returned within 60 minutes of initial draw	Patients hospitalized with ACS **Exclude:** • Patients transferred from a community hospital or ED • Documented decision not to treat within 24 hours. The record clearly documents that the patient, patient's family, or legal representative wishes comfort measures only, and/or there is agreement that the patient's cardiac condition and co-morbid conditions preclude aggressive treatment. Documentation—such as comfort measures only; hospice care; maintain treatment for comfort; terminal care; physician documentation that care is limited at family's

Table D.3—Continued

Measure	Source	Numerator	Denominator
			request or due to patient's age or chronic illness; palliative care; supportive care only— will cause the patient to be excluded from the measure.
Beta Blocker Within 12 hours of Admission	Kerr et al.; 2000b	AMI patients who receive a beta blocker within 12 hours of admission	AMI patients
Stress Test Prior to Discharge	VA, 2007	Patients hospitalized with ACS found to be low-to-moderate-risk ACS patients who received a non-invasive stress test prior to discharge	Patients hospitalized with ACS found to be low-risk ACS patients **Exclude:** • Patients who refuse test
Discharge plan that includes further outpatient stress testing and possible catheterization	VA, 2007	Patients hospitalized with ACS found to be low-risk ACS patients who had a plan prior to discharge that includes further outpatient stress testing and possible catheterization	Patients hospitalized with ACS found to be low-risk ACS patients
Mean Time from Arrival to PCI	JC, 2007a	**Continuous Variable Statement:** Time, in minutes, from hospital arrival to administration of thrombolytic agent in patients with ST-segment elevation or LBBB on the ECG performed closest to hospital arrival time **Excluded Populations:** • Patients less than 18 years of age • Patients received in transfer from another acute care hospital, including another emergency department • Patients with "comfort measures only" documented by a physician, nurse practitioner, or physician assistant • Patients administered fibrinolytic therapy • PCI described as non-primary by physician, nurse practitioner, or physician assistant • Patients who did not receive PCI within 90 minutes and had a reason for delay documented by a physician, nurse practitioner, or physician assistant (e.g., social, religious, initial concern or refusal)	
PCI Received Within 120 Minutes of Hospital Arrival	JC, 2007a; VA, 2007	AMI patients whose time from hospital arrival to PCI is 120 minutes or less	AMI patients with ST segment elevation or LBBB on ECG who received PCI **Exclude:** • Patients less that 18 years of age • Patients received in transfer from another acute

Table D.3—Continued

Measure	Source	Numerator	Denominator
			care hospital, including another ED • Patients admitted on thrombolytic agents
Heparin for at Least 24 Hours	Kerr et al., 2000b	Patients who receive heparin for at least 24 hours	Patients hospitalized with an MI or to rule out MI **Exclude:** • Patients who receive streptokinase, APSAC, or urokinase
Lipid-Lowering Therapy Prescribed at Discharge for Patients with Elevated Low-Density Lipoprotein Cholesterol	JC, 2007a	AMI patients who are prescribed lipid-lowering therapy at hospital discharge	AMI patients with elevated LDL-c **Exclude:** • Patients less than 18 years of age • Patients transferred to another acute care hospital or federal hospital • Patients discharged to hospice • Patients who expired • Patients who left against medical advice • Patients who did not receive lipid-lowering medication and had a reason documented by a physician, nurse practitioner, or physician assistant for not prescribing lipid-lowering medication at discharge
LDL Cholesterol Testing within 24 Hours After Hospital Arrival	JC, 2007a	AMI patients who received LDL-c testing within 24 hours after hospital arrival	AMI patients **Exclude:** • Patients less than 18 years of age • Patients transferred from another acute care hospital or federal hospital on day of arrival • Patients received in transfer from another acute care hospital, including another ED • Patients who expired on day of arrival • Patients who left against medical advice on day of arrival • Patients discharged on day of arrival
Beta Blocker Within 24 Hours of Arrival	JC, 2007a	AMI patients who receive a beta blocker within 24 hours after hospital arrival	AMI patients without beta blocker contraindications

101

Table D.3—Continued

Measure	Source	Numerator	Denominator
			Exclude: • Patients less than 18 years of age • Patients transferred to another acute care hospital or federal hospital on day of arrival • Patients received in transfer from another acute care hospital, including another emergency department • Patients discharged on day of arrival • Patients who expired on day of arrival • Patients who left against medical advice on day of arrival • Patients with one or more of the following contraindications/reasons for not prescribing a beta blocker documented in the medical record: beta blocker allergy; bradycardia on day of discharge or day prior to discharge when not on beta blocker; second- or third-degree heart block on ECG on arrival or during hospital stay and does not have a pacemaker; shock on arrival or within 24 hours after arrival, systolic blood pressure less than 90mm Hg on day of discharge or day prior to discharge while not on beta blocker; other reasons documented by a physician, nurse practitioner, or physician assistant for not prescribing a beta blocker
Aspirin Within 24 Hours Before or After Hospital Arrival	JC, 2007a	AMI patients who received aspirin within 24 hours before or after hospital arrival	AMI patients without aspirin contraindications **Exclude:** • Patients less than 18 years of age • Patients transferred to another acute care hospital or federal hospital on day of arrival • Patients received in transfer from another acute care hospital, including another emergency department • Patients discharged on day of arrival

102

Table D.3—Continued

Measure	Source	Numerator	Denominator
			• Patients who expired on day of arrival • Patients who left against medical advice on day of arrival • Patients with one or more of the following contraindications/reasons for not prescribing aspirin documented in the medical record: active bleeding on arrival or within 24 hours of arrival; aspirin allergy; coumadin/warfarin as pre-arrival medication; other reasons documented by a physician, nurse practitioner, or physician assistant for not giving aspirin within 24 hours before or after hospital arrival
ECG Within 10 Minutes of Arrival	VA, 2007	Patients hospitalized with ACS who had in-hospital ECG performed within 10 minutes of arrival	Patients hospitalized with ACS who before or on arrival at this or another hospital had any cardiac symptoms **Exclude:** • Patients transferred from a community hospital • Patients who are already an inpatient when experiencing an AMI • Documented decision not to treat within 24 hours. The record clearly documents that the patient, patient's family, or legal representative wishes comfort measures only, and/or there is agreement that the patient's cardiac condition and co-morbid conditions preclude aggressive treatment. Documentation—such as comfort measures only; hospice care; maintain treatment for comfort; terminal care; physician documentation that care is limited at family's request or due to patient's age or chronic illness; palliative care; supportive care only—will cause the patient to be excluded from the measure.

Appendix E.
Concepts Suggested by Clinical Advisory Panel

Concept	Description
Cross-Cutting Concepts	
Unnecessary Repetition of Lab Tests	Multiple lab tests of the same type ordered by multiple physicians for a patient because test results were not documented in the patient's chart or the report was not shared with all necessary physicians in a timely manner.
Unnecessary Repetition of CT (Computerized Tomography) Scans	Multiple CT scans ordered by multiple physicians for a patient because results were not documented in patient's chart or the report was not shared with all necessary physicians in a timely manner.
Nursing Turnover	Rate at which nurses voluntarily leave a facility and must be replaced.
Nurse Absenteeism Rates	Rate at which nurses are absent from the facility due to sick time or other personal days.
Labor and Delivery	
None	
Surgery	
Marking the Operative Site (Joint Commission, 2007b)	The unambiguous identification of the intended site of surgical incision or insertion.
"Time Out" Immediately Before Starting Procedure (Joint Commission, 2007b)	A brief period before a surgery is started to conduct a final verification of the correct patient, procedure, site, and, as applicable, implants.
Pre-Op Verification Process (Joint Commission, 2007b)	Completion of a process to ensure that all of the relevant documents and studies are available before the start of the procedure. Missing information or discrepancies should be addressed before starting the procedure.
Formal Briefing Before Surgery (Clinical Advisory Panel)	A brief discussion of the entire surgical team to ensure that everyone is aware of the surgical plan for the patient.
Development of Decubitus Ulcers During Hospital Stay (Clinical Advisory Panel)	Incidence of the development of decubitus ulcers by patients during a hospital stay.
Prevention of Hypothermia (Mangram et al., 1999)	Use of this surgical technique to reduce the risk of surgical-site infection

Perioperative Glucose Control for Diabetics (AHRQ, 2001)	The control of glucose level among diabetic surgical patients.

Acute Myocardial Infarction

None

Appendix F.
Empirical Results for Labor and Delivery Measures

Table F.1. Number of Delivery Discharges in U.S. MTFs, 2002–2004

	2002	2003	2004
Number of MTFs with at least one delivery discharge[a]	70	67	65
Average number of delivery discharges per MTF among MTFs with at least one delivery discharge	669.4	695.3	546.0
Standard deviation for the average number of delivery discharges per MTF	738.4	737.0	586.0
Total number of delivery discharges	46,855	46,587	35,492

[a] Delivery discharges were identified for newborns using DRGs 385–391 and 600–635.

Table F.2. Denominators for Labor and Delivery Outcome Measures

Measure Name	Denominator Definition	2002			2003			2004		
		MTFs with Eligible Patients[a]	Eligible Patients per MTF		MTFs with Eligible Patients[a]	Eligible Patients per MTF		MTFs with Eligible Patients[a]	Eligible Patients per MTF	
			Avg.[b]	Std. Dev.[c]		Avg.[b]	Std. Dev.[c]		Avg.[b]	Std. Dev.[c]
Uterine Rupture (PSTT)	Number of women who delivered infants	63	126.6	123.5	63	124.8	120.0	62	119.9	114.5
Maternal Death (PSTT)	Number of women giving birth at the facility	64	300.7	303.9	64	311.3	314.8	62	322.7	316.8
Intrapartum Fetal Death (term baby) (PSTT)	Number of births at the facility	70	623.1	668.4	67	616.2	627.3	65	495.4	502.9
Birth Trauma, Injury to Neonate Cesarean Section (Subset of PSI 17)	Number of infants delivered via cesarean section	64	155.1	167.8	63	165.5	180.0	62	135.8	143.0
Birth Trauma, Injury to Neonate—Vaginal Delivery (Subset of PSI 17)	Number of infants delivered vaginally	64	546.0	541.2	63	542.5	528.5	63	404.0	417.6

[a] The number of MTFs with at least one patient meeting the criteria for the denominator of the measure.

[b] The average number per MTF of patients meeting the criteria for the denominator of the measure.

[c] Among MTFs with at least one patient meeting the criteria for the denominator of the measure, the standard deviation for the average number per MTF of patients meeting the criteria for the denominator of the measure.

Appendix G.
Empirical Results for Surgery Measures

Table G.1. Number of Surgery Discharges in U.S. MTFs, 2002–2004

	2002	2003	2004
Number of MTFs with at least one surgery discharge[a]	88	85	82
Average number of surgery discharges per MTF among MTFs with at least one surgery discharge	630.0	631.5	676.0
Standard deviation for the average number of surgery discharges per MTF	920.8	928.2	987.9
Total number of surgery discharges	55,444	53,678	55,436

[a] Surgery discharges were identified as admissions with Product Line = Surgery.

Table G.2. Denominators for Surgery Measures[a]

Measure Name	Denominator Definition	2002			2003			2004		
		MTFs With Eligible Patients[b]	Eligible Patients per MTF		MTFs With Eligible Patients[b]	Eligible Patients per MTF		MTFs With Eligible Patients[b]	Eligible Patients per MTF	
			Avg.[c]	Std. Dev.[d]		Avg.[c]	Std. Dev.[d]		Avg.[c]	Std. Dev.[d]
Failure to Rescue (PSI 4)	Number of surgical discharges with potential complications of care listed in *failure to rescue* definition (exclusions listed on page 39)	62	20.1	36.24	64	20.00	37.17	61	21.59	37.24
Foreign Body Left in During Procedure (PSI 5)	Number of surgical discharges defined by specific DRGs.	88	781.0	1079.5	85	811.0	1112.0	82	840.6	1143.1
Mortality in Low Mortality DRG (PSI 2)	Number of surgical patients in DRGs with less than 0.5% mortality rate. If a DRG is divided into "without/with complications," both DRGs must have mortality rates below 0.5% to qualify for inclusion.	88	483.6	599.2	85	517.4	631.7	82	498.3	592.3

[a] Notes: For these measures, the population was limited to surgical patients.

[b] The number of MTFs with at least one patient meeting the criteria for the denominator of the measure.

[c] The average number per MTF of patients meeting the criteria for the denominator of the measure.

[d] Among MTFs with at least one patient meeting the criteria for the denominator of the measure, the standard deviation for the average number per MTF of patients meeting the criteria for the denominator of the measure.

References

AAP (American Academy of Pediatrics). (2007) Hospital stay for healthy term newborns. AHRQ National Guideline Clearinghouse. As of June 4, 2007: http://www.guideline.gov/summary/summary.aspx?doc_id=5088&nbr=003555&string=Healthy+AND+term+AND+newborns. Accessed 6/4/2007.

Adams K and Corrigan JM, eds. Institute of Medicine's Committee on Identifying Priority Areas for Quality Improvement. (2003) *Priority Areas for National Action: Transforming Health Care Quality*. National Academy Press, Washington, DC.

Agency for Healthcare Research and Quality (AHRQ) (2001) Evidence Report/Technology Assessment No. 43, *Making Health Care Safer: A Critical Analysis of Patient Safety Practices*, AHRQ Publication No. 01-E058. Rockville, MD.

(AHRQ) (2005). *Advances in Patient Safety: From Research to Implementation*. Volumes 1–4, AHRQ Publication Nos. 050021 (1–4). February, Rockville, MD.

AHRQ (2007a) Inpatient Quality Indicators. As of June 4, 2007: http://www.qualityindicators.ahrq.gov/iqi_overview.htm.

AHRQ (2007b) AHRQ Patient Safety Indicators. As of June 4, 2007 http://www.qualityindicators.ahrq.gov/psi_overview.htm.

AHRQ (2006) *Guide to Patient Safety Indicators Version 3.0a*. May. Rockville, MD.

Awad SS, Fagan SP, Bellows C, et al. (2005) Bridging the communication gap in the operating room with medical team training. *American Journal of Surgery* 190:770-774.

Bach PB, Cramer LD, Schrag D, et al. (2001) The influence of hospital volume on survival after resection for lung cancer. *New England Journal of Medicine* 345:181-188.

Baggs JG, Ryan SA, Phelps CE, et al. (1992) The association between interdisciplinary collaboration and patient outcomes in a medical intensive care unit. *Heart & Lung* 21:18-24.

Baggs JG, Schmitt MH, Mushlin AI, et al. (1999) Association between nurse-physician collaboration and patient outcomes in three intensive care units. *Critical Care Medicine* 27:1991-1998.

Baker DP, Gustafson S, Beaubien J, et al. (2005a) *Medical Teamwork and Patient Safety: The Evidence-Based Relation*. Literature Review. Agency for Healthcare Research and Quality, AHRQ Publication No. 05-0053, April 2005. Rockville MD. As of September, 29, 2005: http://www.ahrq.gov/qual/medteam.

Baker DP, Salas E, King H, et al. (2005b) The role of teamwork in professional education of physicians: Current status and assessment recommendations. *Joint Commission Journal on Quality and Patient Safety* 31:185-202.

Barrett J, Gifford C, Morey J, et al. (2001) Enhancing patient safety through teamwork training. *Journal of Healthcare Risk Management.* 21:57-65.

Best WR, Khuri SF, Phelan M, et al. (2002) Identifying patient preoperative risk factors and postoperative adverse events in administrative databases: Results from the Department of Veterans Affairs National Surgical Quality Improvement Program. *Journal of the American College of Surgeons* 194:257-266.

Blum RH, Raemer DB, Carroll JS, et al. (2004) Crisis resource management training for an anesthesia faculty: A new approach to continuing education. *Medical Education* 38:45-55.

Brennan TA, Leape LL, Laird NM, et al. (1991) Incidence of adverse events and negligence in hospitalized patients: Results of the Harvard Medical Practice Study I, *New England Journal of Medicine* 324:370-376.

Boyle SM. (2004) Nursing unit characteristics and patient outcomes. *Nursing Economics* 22:111-123.

Campbell SM, Hann M, Hacker J, et al. (2001) Identifying predictors of high quality care in English general practice: Observational study. *BMJ* 323:784-792.

Cannon-Bowers JA, Tannenbaum SI, Salas E, et al. (1995) Defining competencies and establishing team training requirements. In: Guzzo RA, Salas E, and Associates, eds. *Team Effectiveness and Decision-Making in Organizations.* Josey-Bass, San Francisco, CA,: pp 333-380.

Carthey J, de Leval MR, Wright DJ, et al. (2003) Behavioral markers of surgical excellence. *Safety Science* 41:409-425.

Cashman SB, Reidy P, Cody K, Lemay C. (2004) Developing and measuring progress toward collaborative, integrated, interdisciplinary health care teams. *Journal of Interprofessional Care* 18:183-196.

CMS (Centers for Medicare and Medicaid Services). The Premier Hospital Quality Demonstration: Clinical Conditions and Measures for Reporting. As of March 31, 2005: http://www.cms.hhs.gov/HospitalQualityInits/downloads/HospitalPremierMeasures.pdf.

Curley C, McEachern JE, Speroff T. (1998) A firm trial of interdisciplinary rounds on the inpatient medical wards: An intervention designed using continuous quality improvement. *Medical Care* 36(2 Supplement):AS4-AS12.

Deeter-Schmelz D, Kennedy KN. (2003) Patient care teams and customer satisfaction: The role of team cohesion. *Journal of Services Marketing* 17:666-682.

DeFones J, Surbida S. (2004) Preoperative safety briefing project. *The Permanente Journal* 8:21-27.

DeVita MA, Schaefer J, Lutz J, Wang H, Dongilli T. (2005) Improving medical emergency team (MET) performance using a novel curriculum and a computerized human patient simulator. *Quality and Safety in Health Care* 14:326-331.

Dienst ER, Byl N. (1981) Evaluation of an education program in health care teams. *Journal of Community Health* 6:282-298.

Donchin Y, Gopher D, Olin M, et al. (1995) A look into the nature and causes of human errors in the intensive care unit. *Critical Care Medicine* 23:294-300.

Edmondson A, Bohmer R, Pisano G. (2001) Speeding up team learning. *Harvard Business Review* October:5-11.

El-Jardali F, Lagace M. (2005) Making hospital care safer and better: The structure-process connection leading to adverse events. *Healthcare Quarterly* 8(2):40-48.

Faciszewski T, Smith NC. (1995) Administrative databases' complication coding in anterior spinal fusion procedures. What does it mean? *Spine* 20:1783-1788.

Feinglass J, Koo S, Koh J. (2004) Revision total knee arthroplasty complication rates in Northern Illinois. *Clinical Orthopaedics and Related Research* 429:279-285.

Flanagan B, Nestel D, Joseph M. (2004) Making patient safety the focus: Crisis resource management in the undergraduate curriculum. *Medical Education* 38:56-66.

Flin R, Maran N. (2004) Identifying and training non-technical skills for teams in acute medicine. *Quality and Safety in Health Care* 13:i80-i84.

Friedman DM, Berger DL. (2004) Improving team structure and communication: A key to hospital efficiency. *Archives of Surgery* 139:1194-1198.

Gaba DM, Howard SK, Fish KJ, et al. (2001) Simulation-based training in anesthesia crisis resource management (ACRM): A decade of experience. *Simulation & Gaming* 32:175-193.

Gaba DM, Howard SK, Flanagan B, et al. (1998) Assessment of clinical performance during simulated crisis using both technical and behavioral ratings. *Anesthesiology* 89:3-18.

Geraci JM, Ashton CM, Kuykendall DH, et al. (1997) International Classification of Diseases, 9th Revision, Clinical Modification Codes in discharge abstracts are poor measures of complication occurrence in medical inpatients. *Medical Care* 35: 589-602.

Gitell JH, Fairfield KM, Bierbaum B, et al. (2000) Impact of relational coordination on quality of care, postoperative pain and functioning, and length of stay: A nine hospital study of surgical patients. *Medical Care* 38:807-819.

Glance LG, Dick AW, Osler TM, Mukamel DB. (2006) Accuracy of hospital report cards based on administrative data. *HSR: Health Services Research* 41:1413-1437.

Goni S. (1999) An analysis of the effectiveness of Spanish primary health care teams. *Health Policy* 48:107-117.

Grobman WA, Feinglass J, Murthy S. (2006) Are the Agency for Healthcare Research and Quality obstetric trauma indicators valid measures of hospital safety? *American Journal of Obstetrics and Gynecology* 195:868-874.

Grogan EL, Stiles RA, France DJ, et al. (2004) The impact of aviation-based teamwork training on the attitudes of health-care professionals. *Journal of the American College of Surgeons* 199:843-848.

Haig KM, Sutton S, Whittington J. (2006) SBAR: A shared mental model for communication between clinicians. *Joint Commission Journal on Quality and Patient Safety* 32:167-175.

Halamek LP, Kaegi DM, Gaba DM, et al. (2000) Time for a new paradigm in pediatric medical education: Teaching neonatal resuscitation in a simulated delivery room environment. *Pediatrics* 106:e45-e50.

Hamman WR. (2004) The complexity of team training: What we have learned from aviation and its applications to medicine. *Quality and Safety in Health Care* 13:i72-i79.

Hawker GA, Coyte PC, Wright JG, et al. (1997) Accuracy of administrative data for assessing outcomes after knee replacement surgery. *Journal of Clinical Epidemiology* 50:265-273.

Healey AN, Undre S, Vincent CA. (2004) Developing observational measures of performance in surgical teams. *Quality and Safety in Health Care* 13(Supple 1):i33-i40.

Helmreich RL. (2000) On error management: Lessons from aviation. *British Medical Journal* 320:781-785.

Helmreich RL. (1987) Theory underlying CRM training: Psychological issues in flight crew performance and crew coordination. In: Orlady HW, Foushee HC, eds. *Cockpit resource management training: Proceedings of the NASA/MAC Workshop.* NASA-Ames Research Center, CP-2455, Moffett Field, CA.

Helmreich RL, Schaefer HG. (1994) Team performance in the operating room. In MS Bogner, ed. *Human Error in Medicine.* Lawrence Erlbaum Associates, Hillsdale, N J.

Hicks RW, Dantell JP, Cousins DD, et al. (2004) *MEDMARX^{SM} 5th Anniversary Data Report: A Chartbook of 2003 Findings and Trends 1999-2003.* USP Center for the Advancement of Patient Safety, Rockville, MD.

Holzman RS, Cooper JB, Gaba DM, et al. (1995) Anesthesia crisis resource management: Real-life simulation training in operating room crises. *Journal of Clinical Anesthesia* 7:675-687.

Hope JM, Lugassy D, Meyer R, et al. (2005) Bringing interdisciplinary and multicultural team building to health care education: The Downstate Team-Building Initiative. *Academic Medicine* 80:74-83.

Horak BJ, Pauig J, Keidan B, Kerns J. (2004) Patient safety: A case study in team building and interdisciplinary collaboration. *Journal for Healthcare Quality* 26(2):6-13.

Howard SK, Gaba DM, Fish KJ, et al. (1992) Anesthesia crisis resource management training: teaching anesthesiologists to handle critical incidents. *Aviation Space & Environmental Medicine* 63:763-770.

Hospital Compare Web Site. As of November 19, 2007: http://www.hospitalcompare.hss/gov/hospitalsearch/searchcriteria.asp?version=default& browsersafari%7c2%7Cmacosx&language=english=defaultstatus=0&pagelist=home

i Gardi T, Christensen UC, Jacobsen J, et al. (2001) How do anaesthesiologists treat malignant hyperthermia in a full-scale anaesthesia simulator? *Acta Anaesthesiologica Scandinavica* 45:1032-1035.

ICSI (Institute for Clinical Systems Improvement). (2007) AHRQ National Quality Measures Clearinghouse. As of June 5, 2007: http://www.qualitymeasures.ahrq.gov/.

Iezzoni LI, (ed.) (1994) *Risk Adjustment for Measuring Health Care Outcomes.* Health Administration Press, Ann Arbor, MI.

Institute of Medicine, Committee on Quality of Health Care in America. (2001) *Crossing the Quality Chasm: A New Health System for the 21st Century.* National Academy Press, Washington, DC.

Jacobs R, Goddard M, Smith PC. (2005) How robust are hospital ranks based on composite performance measures? *Medical Care* 43:1177-1184.

Jacobsen J, Lindekaer AL, Ostergaard HT, et al. (2001) Management of anaphylactic shock evaluated using a full-scale anaesthesia simulator. *Acta Anaesthesiologica Scandinavica* 45:315-319.

Johnson WG, Brennan TA, Newhouse JP, et al. (1992) The Economic Consequences of Medical Injuries. *JAMA* 267:2487-2492.

Joint Commission on Accreditation of Healthcare Organizations. (2004a) *Sentinel Event Statistics.* As of December 27, 2004: www.jcaho.org/accredited+organizations/ambulatory+care/ sentinel+events.htm. As of November 26, 2007: http://www.jointcommission.org/sentinelevents/statistics/

Joint Commission on Accreditation of Healthcare Organizations. (2004b) *National Patient Safety Goals for 2004 and 2005.* As of December 27, 2004: www.jcaho.org/accredited+organizations/ patient+safety/npsg.htm..

Joint Commission on Accreditation of Healthcare Organizations (Joint Commission). (2004c) *Preventing Infant Death During Delivery.* Sentinel Event Alert No. 30. As of November 26, 2007: http://www.jointcommission.org/SentinelEvents/SentinelEventAlert/sea_30.htm.

Joint Commission. (2006) *2006 National Patient Safety Goals Slideshow.* As of May 22, 2006: http://www.jointcommission.org/NR/rdonlyres/88ED0EA4-395A-4DBA-A12E-25C70AF72C00/0/06_npsg.ppt.

Joint Commission. (2007a) Joint Commission Core Measures. As of June 5, 2007: http://www.jointcommission.org/PerformanceMeasurement/PerformanceMeasurement/default.htm.
As of November 26, 2007: http://www.jointcommission.org/Performance/Measurement/PerformanceMeasurement/current+NHQM+manual.htm.
Joint Commission. (2007b) Joint Commission Protocols. As of June 5, 2007: http://www.jointcommission.org/NR/rdonlyres/E3C600EB-043B-4E86-B04E-CA4A89AD5433/0/universal_protocol.pdf. .

Joint Commission. (2007c) Joint Commission Recommended Measures for Future Implementation. As of June 5, 2007 http://www.jointcommission.org/NR/rdonlyres/48DFC95A-9C05-4A44-AB051769D5253014/0/AComprehensiveReviewofDevelopmentforCoreMeasures.pdf

Kerr EA, Asch SM, Hamilton EG, et al. eds. (2000a) *Quality-of-care for General Medical Conditions: A Review of the Literature and Quality Indicators,* RAND Corporation, Santa Monica, CA, MR-1280-AHRQ, As of November 26, 2007: http://www.rand.org/publications/MR/MR1280.

Kerr EA, Asch SM, Hamilton EG, et al. eds. (2000b) *Quality-of-care for Cardiopulmonary Conditions: A Review of the Literature and Quality Indicators,* RAND Corporation, Santa Monica, CA, MR-1282-AHRQ, As of November 26, 2007: http://www.rand.org/publications/MR/MR1282.

Kivimaki M, Sutinen R, Elovainio M, et al. (2001) Sickness absence in hospital physicians: 2 year follow up study on determinants. *Occupational and Environmental Medicine* 58:361-366.

Knaus WA, Draper EA, Wagner DP, Zimmerman JE. (1986) An evaluation of outcome from intensive care in major medical centers. *Annals of Internal Medicine* 104:410-418.

Knaus WA, Wagner DP, Draper EA, et al. (1991) The APACHE III prognostic system: Risk prediction of hospital mortality for critically ill hospitalized adults. *Chest* 100:1619-1636.

Koerner BL, Cohen JR, Armstrong DM. (1985) Collaborative practice and patient satisfaction: Impact and selected outcomes. *Evaluation & the Health Professions* 8:299-321.

Kohn LT, Corrigan JM, Donaldson MS, eds. (2000) *To Err Is Human: Building a Safer Health System.* National Academy Press, Washington, DC.

Kurrek MM, Fish KJ. (1996) Anaesthesia crisis resource management training: An intimidating concept, a rewarding experience. *Canadian Journal of Anaesthesia* 43:430-434.

Landro L. (2005) Bringing surgeons down to Earth: New programs aim to curb fear that prevents nurses from flagging problems. *The Wall Street Journal* November 16.

Leonard M, Graham S, Bonacum D. (2004) The human factor: The critical importance of effective teamwork and communication in providing safe care. *Quality and Safety in Health Care* 13:i85-i90.

MacLean CH, Louie R, Shekelle PG, et al. (2006) Comparison of administrative data and medical records to measure the quality of medical care provided to vulnerable older patients. *Medical Care* 44:141-148.

Makary M, Sexton JB, Freischlag J, et al. (2006) Operating room teamwork among physicians and nurses: Teamwork in the eye of the beholder. *Journal of the American College of Surgeons* 202:746-752.

Mangram A, Horan T, Pearson M, et al. (1999) guideline for prevention of surgical site infection. *Infection Control and Hospital Epidemiology* 20:247-278.

Mann S, Marcus R, Sachs B. (2006) Lessons from the cockpit: How team training can reduce errors on L&D. *Contemporary Ob/Gyn* 51:34-45.

Mann S, Pratt S, Gluck P et al. (2006) Assessing quality in obstetrical care: Development of standardized measures. *Joint Commission Journal on Quality and Patient Safety* 32:497-505.

McCarthy EP, Iezzoni LI, Davis RB, et al. (2000) Does clinical evidence support ICD-9-CM diagnosis coding of complications? *Medical Care* 38:868-876.

McGlynn E, Kerr EA, Damberg CL, et al. eds. (2000) *Quality-of-care for Women: A Review of the Selected Clinical Conditions and Quality Indicators,* RAND Corporation, Santa Monica, CA, MR-1284-HCFA: http://www.rand.org/publications/MR/MR1284.

Mehrotra A, Bodenheimer T, Dudley RA. (2003) Employers' efforts to measure and improve hospital quality: Determinants of success. *Health Affairs* 22:60-71.

Meterko M, Mohr DC, Young GJ. (2004) Teamwork culture and patient satisfaction in hospitals. *Medical Care* 42:492-498.

Military Health System, (2007) *Military Health System Strategic Plan.* As of June 6, 2007: http://www.ha.osd.mil/strat_plan/MHS_Strategic_Plan_07Apr.pdf.

Miller MR, Elixhauser A, Zhan C, Meyer GS. (2001) Patient Safety Indicators: Using administrative data to identify potential patient safety concerns. *HSR: Health Services Research* 36(6 pt 2):110-132.

Morey JC, Simon R, Jay GD, et al. (2002) Error reduction and performance improvement in the emergency department through formal teamwork training: Evaluation results of the MedTeams project. *Health Services Research* 37:1553-1581.

Mukamel DB, Temkin-Greener H, Delavan R, et al. (2006) Team performance and risk-adjusted health outcomes in the Program of All-Inclusive Care for the Elderly (PACE). *The Gerontologist* 46:227-237.

Murff HJ, Patel VL, Hripcsak G, Bates D. (2003) Detecting adverse events for patient safety research: A review of current methodologies. *Journal of Biomedical Informatics* 36:131-143.

National Quality Forum. (2006) Safe practices for better healthcare —2006 update. As of April 8, 2006:
http://www.qualityforum.org/pdf/projects/safe-practices/AppealsDraft-Background-and-UpdatedSafePractices.pdf

NQF (National Quality Forum). (2007) Society for Thoracic Surgeons/NQF National Voluntary Consensus Standards for Cardiac Surgery: (2004) A Consensus Report, As of June 4, 2007: http://216.122.138.39/pdf/reports/cardiac.pdf.

O'Connell MT, Pascoe JM (2004) Undergraduate medical education for the 21st century: Leadership and teamwork. *Family Medicine* 36:S51-S56.

O'Donnell J, Fletcher J, Dixon B, et al. (1998) Planning and implementing an anesthesia crisis resource management training course for student nurse anesthetists. *CRNA* 9:50-58.

Ostergaard HT, Ostergaard D, Lippert A. (2004) Implementation of team training in medical education in Denmark. *Quality and Safety in Health Care* 13:i91-i95.

Persell SD, Wright JM, Thompson JA, et al. (2006) Assessing the validity of national quality measures for coronary artery disease using an electronic health record. *Archives of Internal Medicine* 166:2272-2277.

Pham HH, Coughlan J, O'Malley AS. (2006) The impact of quality reporting programs on hospital operations. *Health Affairs* 25:1412-1422.

Pronovost P, Berenholtz S, Dorman T, et al. (2003) Improving communication in the ICU using daily goals. *Journal of Critical Care* 18(2):71-75.

Pronovost P, Needham D, Berenholtz S, et al. (2006) An intervention to decrease catheter-related bloodstream infections in the ICU. *The New England Journal of Medicine* 355:2725-2732.

Pronovost PJ, Wu AW, Sexton JB. (2004) Acute decomposition after removing a central line: Practical approaches to increasing safety in the intensive care unit. *Annals of Internal Medicine* 140:1025-1033.

Quality Interagency Coordination Task Force. (2000) *Doing What Counts for Patient Safety: Federal Actions to Reduce Medical Errors and Their Impact.* Report of the Quality Interagency Coordination Task Force (QuIC) to the President. Rockville, MD.

Rafferty AM, Ball J, Aiken LH. (2001) Are teamwork and professional autonomy compatible, and do they result in improved hospital care? *Quality in Health Care* 10:ii32-ii37.

Remus D, Fraser I. (2004) Guidance for using the AHRQ quality indicators for hospital-level public reporting or payment. Agency for Health Care Research and Quality, AHRQ Publication No.04-0086-EF, Rockville, MD.

Resar RK, Rozich JD, Classen D. (2003) Methodology and rationale for the measurement of harm with trigger tools. *Quality and Safety in Health Care* 12: ii39-ii45.

Reznek M, Smith-Coggins R, Howard S, et al. (2003) Emergency medicine crisis resource management (EMCRM): A pilot study of a simulation-based crisis management course for emergency medicine. *Academic Emergency Medicine* 10:386-389.

Risser DT, Rice MM, Salisbury ML, et al. (1999) The potential for improved teamwork to reduce medical errors in the emergency department. *Annals of Emergency Medicine* 34:373-383.

Risser D, Barrett J, et al. (2002) MedTeams: Labor and Delivery Measures Guide. Dynamics Research Corporation, Andover, MA.

Rivers RM, Swain D, Nixon WR. (2003) Using aviation safety measures to enhance patient outcomes. *AORN Journal* 77(1):158-162.

Romano PS, Chan BK, Schembri ME, Rainwater JA. (2002) Can administrative data be used to compare postoperative complication rates across hospitals? *Medical Care* 40:856-867.

Romano PS, Schembri ME, Rainwater JA. (2002) Can administrative data be used to ascertain clinically significant postoperative complications? *American Journal of Medical Quality* 17:145-154.

Rosen AK, Zhao S, Rivard P, et al. (2006) Tracking rates of patient safety indicators over time: Lessons from the Veterans Administration. *Medical Care* 44:850-861.

Rosenthal S, Chen R. (1995) The reporting sensitivities of two passive surveillance systems for vaccine adverse events. *The American Journal of Public Health* 85:1706-1709.

Rozich JD, Haraden CR, Resar RK. (2003) Adverse drug event trigger tool: A practical methodology for measuring medication-related harm. *Quality and Safety in Health Care* 12:194-200.

Salas E, Burke CS, Bowers CA, et al. (2001) Team training in the skies: Does crew resource management (CRM) training work? *Human Factors* 43: 641-674.

Schenkel S. (2000) Promoting patient safety and preventing medical error in emergency departments. *Academic Emergency Medicine* 7:1204-1222.

National Cancer Institute, SEER (Surveillance, Epidemiology and End Results Web site). (2007) As of November 26, 2007: http://seer.cancer.gov/.

Sexton JB, Thomas EJ, Helmreich RL. (2000) Error, stress, and teamwork in medicine and aviation: Cross sectional surveys. *British Medical Journal* 320:745-749.

Shapiro MJ, Morey JC, Small SD, et al. (2004) Simulation based teamwork training for emergency department staff: Does it improve clinical team performance when added to an existing didactic teamwork curriculum? *Quality and Safety in Health Care* 13:417-421.

Shojania KG, Duncan BW, McDonald KM, Wachter RM, eds. (2001) *Making Health Care Safer: A Critical Analysis of Patient Safety Practices.* Agency for Healthcare Research and Quality, Evidence Report/Technology Assessment #43, AHRQ Publication No. 01-E058, Rockville, MD.

Shortell SM, Jones R, Rademaker A, et al. (2000) Assessing the impact of total quality management and organizational culture on multiple outcomes of care for coronary artery bypass graft surgery patients. *Medical Care* 38:207-217.

Shortell SM, Zimmerman JE, Rousseau DM, et al. (1994). The performance of intensive care units: Does good management make a difference? *Medical Care* 32:508-525.

Sica GT, Barron DM, Blum R, et al. (1999) Computerized realistic simulation: A teaching module for crisis management in radiology. *American Journal of Roentgenology* 172:301-304.

Skinner KA, Helsper JT, Deapen, D et al. (2003) Breast cancer: Do specialists make a difference? *Annals of Surgical Oncology* 10:606-615.

Stevenson K, Baker R, Farooqi A, et al. (2001) Features of primary health care teams associated with successful quality improvement of diabetes care: A qualitative study. *Family Practice* 18(1):21-26.

Stoller JK, Rose M, Lee R, et al. (2004) Teambuilding and leadership training in an internal medicine residency training program: Experience with a one-day retreat. *Journal of General Internal Medicine* 19:692-697.

Surgical Care Improvement Project (SCIP) (2007) Project Measures. As of June 5, 2007: http://www.medqic.org/dcs/ContentServer?cid=1122904930422&pagename=Medqic%2FMe asure%2FMeasuresHome&parentName=Topic&level3=Measures&c=MQParents.

Taub DA, Miller DC, Cowan JA, et al. (2004) Impact of surgical volume on mortality and length of stay after nephrectomy. *Urology* 63:862-867.

TeamSTEPPS™ Multimedia Resource Kit. (2006) [TeamSTEPPS™: Team Strategies and Tools to Enhance Performance and Patient Safety; developed by the Department of Defense and published by the Agency for Healthcare Research and Quality.] Agency for Healthcare Research and Quality, AHRQ Publication No. 06-0020-4, Rockville, MD.

Thomas EJ, Peterson L, (2003) Measuring errors and adverse events in health care. *Journal of General Internal Medicine* 18:61-67.

Thomas EJ, Sexton JB, Helmreich RL. (2003) Discrepant attitudes about teamwork among critical care nurses and physicians. *Critical Care Medicine* 31:956-959.

Thomas EJ, Sexton JB, Helmreich RL. (2004) Translating teamwork behaviors from aviation to healthcare: Development of behavioral markers for neonatal resuscitation. *Quality and Safety in Health Care* 13:i57-i64.

Thomas EJ, Sexton JB, Lasky RE, et al. (2006) Teamwork and quality during neonatal care in the delivery room. *Journal of Perinatology* 26:163-169.

Thomas EJ, Studdert DM, Newhouse JP, et al. (1999) Costs of medical injuries in Utah and Colorado. *Inquiry* 36:255-264.

Uhlig PN, Brown J, Nason AK, et al. (2002) System innovation: Concord Hospital. *Joint Commission Journal on Quality Improvement* 28:666-672.

Undre S, Healy AN, Darzi A, Vincent CA. (2006) Observational assessment of surgical teamwork: A feasibility study. *World Journal of Surgery* 30:1774-1783.

US Pharmacopeia. (2003a) Miscommunication leads to confusion and errors. *CAPSLink* December.

US Pharmacopeia (2003b). Medication errors in the perioperative environment. *CAPSLink* March.

VA (Veterans Health Administration). (2007) Veterans Health Administration Performance Measures. AHRQ National Quality Measures Clearinghouse. As of June 6, 2007: http://www.qualitymeasures.ahrq.gov/.

Wheelan SA, Burchill CN, Tilin F. (2003) The link between teamwork and patients' outcomes in intensive care units. *American Journal of Critical Care* 12:527-534.

White A, Pichert J, Bledsoe S, et al. (2005) Cause and effect analysis of closed claims in obstetrics and gynecology. *Obstetrics and Gynecology* 105:1031-1038.

Young MP, Gooder VJ, Oltermann MH, et al. (1998) The impact of a multidisciplinary approach on caring for ventilator-dependent patients. *International Journal for Quality in Health Care* 10:15-26.